Reviews of *3G Marketing*:

"*3G Marketing* is packed with useful and practical techniques and examples of how mobile operators can use modern targeted marketing to achieving success in the competitive marketplace."

Steven S K Chan
Director, Internet Services and Product Development
MobileOne, Singapore

"The authors have accurately described the issues involved in the introduction of radically new services into untested markets, and provide a wealth of practical tools and methods to help achieve market success."

Claus Nehmzov
International Development Director
Shazam Entertainment, UK

"In this book the authors offer an insightful look into how modern wireless carriers are capitalising on their customer data and developing targeted marketing propositions."

Jan-Anders Dalenstam
Sr Vice President, Business Development
Ericsson Wireless Communications, USA

"The authors combine a solid marketing foundation with the latest mobile telecoms phenomena such as reachability and communities to create a handbook for achieving customer satisfaction in the connected age."

Alan Moore
CEO
Small Medium Large Xtralarge (SMLXL), UK

"This book is the first to explain how telecoms billing, tariffing, and revenue assurance all relate to profitability as operators rush to deploy advanced and complex services."

David Leshem
Executive Vice President Marketing
Compwise, Israel

3G Marketing

3G Marketing

Communities and Strategic Partnerships

Tomi T Ahonen
Independent Consultant, UK

Timo Kasper
Observer Finland, Finland

Sara Melkko
Independent Expert, Germany

John Wiley & Sons, Ltd

Copyright © 2004   John Wiley & Sons Ltd, The Atrium, Southern Gate,
                     Chichester, West Sussex PO19 8SQ, England

                     Telephone   (+44) 1243 779777

Email (for orders and customer service enquiries): cs-books@wiley.co.uk
Visit our Home Page on www.wileyeurope.com or www.wiley.com

All Rights Reserved. No part of this publication may be reproduced, stored in a retrieval system or transmitted in any form or by any means, electronic, mechanical, photocopying, recording, scanning or otherwise, except under the terms of the Copyright, Designs and Patents Act 1988 or under the terms of a licence issued by the Copyright Licensing Agency Ltd, 90 Tottenham Court Road, London W1T 4LP, UK, without the permission in writing of the Publisher. Requests to the Publisher should be addressed to the Permissions Department, John Wiley & Sons Ltd, The Atrium, Southern Gate, Chichester, West Sussex PO19 8SQ, England, or emailed to permreq@wiley.co.uk, or faxed to (+44) 1243 770620.

This publication is designed to provide accurate and authoritative information in regard to the subject matter covered. It is sold on the understanding that the Publisher is not engaged in rendering professional services. If professional advice or other expert assistance is required, the services of a competent professional should be sought.

Other Wiley Editorial Offices

John Wiley & Sons Inc., 111 River Street, Hoboken, NJ 07030, USA

Jossey-Bass, 989 Market Street, San Francisco, CA 94103-1741, USA

Wiley-VCH Verlag GmbH, Boschstr. 12, D-69469 Weinheim, Germany

John Wiley & Sons Australia Ltd, 33 Park Road, Milton, Queensland 4064, Australia

John Wiley & Sons (Asia) Pte Ltd, 2 Clementi Loop #02-01, Jin Xing Distripark, Singapore 129809

John Wiley & Sons Canada Ltd, 22 Worcester Road, Etobicoke, Ontario, Canada M9W 1L1

Wiley also publishes its books in a variety of electronic formats. Some content that appears in print may not be available in electronic books.

Library of Congress Cataloging in Publication Data

Ahonen, Tomi T.
3G marketing : communities and strategic partnerships / Tomi T. Ahonen, Timo Kasper, Sara Melkko.
    p. cm.
Includes bibliographical references and index.
ISBN 0-470-85100-7 (cloth : alk. paper)
1. Cellular telephone services industry. 2. Cellular telephone equipment industry.
3. Cellular telephones—Marketing. I. Title: Three G marketing. II. Kasper, Timo.
III. Melkko, Sara. IV. Title.
HE9713.A356 2004
384.5′3′0688—dc22   2004011077

British Library Cataloguing in Publication Data

A catalogue record for this book is available from the British Library

ISBN 0-470-85100-7

Typeset in 10/12pt Times by Integra Software Services Pvt. Ltd, Pondicherry, India.
Printed and bound in Great Britain by TJ International Ltd, Padstow, Cornwall.
This book is printed on acid-free paper responsibly manufactured from sustainable forestry in which at least two trees are planted for each one used for paper production.

Contents

About the Authors	xvii
Foreword	xix
Acknowledgements	xxi

1	**Introduction**	**1**
	1.1 A look back	2
	1.1.1 Enter I-Mode	5
	1.1.2 The flap about WAP being a failure	6
	1.1.3 Growth rate	7
	1.2 What have we learned?	8
	1.2.1 Telecoms operators and 3G marketing	9
	1.3 Lets touch upon definitions of 3G	10
	1.3.1 So what is 4G	12
	1.3.2 W-LAN or Wi-Fi is definitely not 4G	12
	1.3.3 4G will arrive ten years from now	13
	1.4 To sum up	13
2	**Market Intelligence**	**15**
	2.1 What is market intelligence	16
	2.1.1 Evolution of market intelligence	16
	2.1.2 Information, analysis, knowledge and intelligence	16
	2.1.3 Knowledge or Intelligence	17

2.2	Systematic market intelligence	17
	2.2.1 Market intelligence and business intelligence	18
	2.2.2 Legal and regulatory intelligence	18
	2.2.3 Customer intelligence	19
	2.2.4 Competitor intelligence	20
	2.2.5 Technical environment intelligence	21
	2.2.6 Telecoms is not used to rapid innovation	22
	2.2.7 The computer industry thrives on rapid innovation	23
2.3	'Environment scanning' intelligence	25
	2.3.1 Resource market intelligence	26
	2.3.2 Reference market studies	27
	2.3.3 Partnership intelligence/networking	27
2.4	Towards a higher intelligence	29

3 Segmentation 31

3.1	What is segmentation?	32
	3.1.1 Test of current telecoms segmentation	33
	3.1.1.1 Segmentation by size	34
	3.1.1.2 Segmentation by technology	34
	3.1.1.3 Segmentation by billing	34
3.2	Segmentation criteria	35
	3.2.1 Segmentation from the academics	35
	3.2.2 Segmentation by geographical pattern	36
	3.2.3 Segment by demographics	36
	3.2.4 Industry type	36
	3.2.5 Segmentation by using various distribution channels	38
	3.2.6 Personal data	38
	3.2.7 Segmentation by psychological patterns	38
3.3	ERP, CRM and segmentation	38
	3.3.1 From hard to soft facts	39
	3.3.2 Users broken down — segmenting situations	40
3.4	From theory to practice: building a segmentation model	41
	3.4.1 Characteristics of a useful segmentation model	41
	3.4.2 Segmentation by user behaviour	41
	3.4.3 How many segments?	43
	3.4.4 Comparison with the car industry	46
	3.4.5 Beyond a segment of one	47
	3.4.6 From business to individual	48
	3.4.7 Self-organizing maps	48
	3.4.8 From alphas to omegas	49
3.5	Developing the segmentation model	51
3.6	To sum up segmentation	54

4	**Service Development and Management**	**55**
	4.1 Product development — the Five Ms	56
	4.1.1 Power of personalization	57
	4.1.2 Money brings content	58
	4.1.3 Talking machines	59
	4.2 Service management (product management)	61
	4.2.1 Knowing the market	62
	4.2.2 New service ideas	62
	4.2.3 Brainstorming	63
	4.2.4 From idea to opportunity	64
	4.2.5 Let there be light	64
	4.2.6 It is your own sales who knows your customer best	65
	4.2.7 Caught in the middle of the triangle	66
	4.3 The launch	67
	4.3.1 Tariffing, cost and profit	67
	4.4 Killing a service	68
	4.5 To finish with service creation	68
5	**Partnership Management**	**71**
	5.1 What is partnering?	72
	5.1.1 Flavours of partnering	73
	5.1.2 Who are the prospective partners?	74
	5.2 Operators are new to this game	75
	5.2.1 Culture shock	76
	5.3 Revenue sharing	78
	5.3.1 What kind of revenue (and/or cost) sharing options?	79
	5.3.2 What level of revenue sharing	80
	5.4 Main factors influencing split in revenue share	81
	5.4.1 Exclusivity	82
	5.4.2 Value chain	82
	5.2.3 On-screen location	83
	5.2.4 Brand strength	83
	5.2.5 Location information	84
	5.2.6 Charging/billing information	84
	5.3 Rules of thumb	84
	5.4 Contract management	86
	5.4.1 Keys to success	87
	5.4.2 Partnering for profit	88
	5.5 Parting with partnering	89
6	**Terminals**	**91**
	6.1 How our gadgets evolve	92
	6.1.1 Convergence	93

	6.2	The Swiss knife or all-in-one device	95
	6.3	Custom-use devices	96
		6.3.1 The PDA	97
		6.3.2 Digital camera	99
		6.3.3 Gaming devices	101
		6.3.4 The credit card	102
		6.3.5 GPS devices	103
		6.3.6 3G modems	103
		6.3.7 Custom devices	103
	6.4	Automobiles	104
		6.4.1 Servicing and maintaining the car	104
		6.4.2 Navigation	105
		6.4.3 Car security and anti-theft	105
		6.4.4 Multitasking and the car	106
		6.4.5 Games in the car	106
	6.5	More devices that seem like science fiction	106
	6.6	Handset subsidies	108
		6.6.1 Device needs	109
		6.6.2 Connectivity	109
		6.6.3 Synchronization	110
	6.7	Handing off on handsets	110
7	**Distribution**		**113**
	7.1	Sales channels	114
		7.1.1 Operator's own stores	114
		7.1.2 Independent stores	115
		7.1.3 Departments and sales desks of other stores	116
		7.1.4 IT integrators	116
		7.1.5 The Internet as a sales channel	117
		7.1.6 The mobile portal as a sales channel	118
		7.1.7 MVNOs	118
	7.2	Managing channel conflicts	118
	7.3	Selling new mobile services	119
		7.3.1 Bundling an m-component	119
		7.3.2 Soul of the store sales rep	120
	7.4	Information flow	121
	7.5	Warehousing, shipping, inventory	122
	7.6	Distribution as an end	123
8	**Portals**		**125**
	8.1	Defining portals	126
	8.2	3G portal categorization	126
		8.2.1 Different types of mobile portal	126
		8.2.2 Categorization	127

8.3		The 3D rule for mobile portals	127
	8.3.1	What is murfing	128
8.4		Personalization	129
	8.4.1	Authentication ('intelligent' portal)	129
	8.4.2	Timing ('instant' portal)	130
	8.4.3	Positioning (portal 'to go')	130
	8.4.4	Pull versus push (portal 'on demand')	130
8.5		Open content policy – a decisive battle over 3G's success	131
	8.5.1	The more services, the more money for everybody	131
	8.5.2	Price strategies: skimming versus penetration	132
8.6		Revenues and advertising	133
8.7		Collect customer data (registration)	133
	8.7.1	Advertising	134
	8.7.2	Buy your ad on the top of search engines	135
	8.7.3	Cross selling (own products)	136
	8.7.4	Customer loyalty programmes/clubs	136
	8.7.5	m-Commerce (partner marketing)	136
	8.7.6	Multi-access portal	136
8.8		Closing the portal	137

9 Promotion   139

9.1		Is the classic marketing mix all mixed up in 3G?	140
	9.1.1	The AIDA rule	140
9.2		Crossing the 3G chasm	141
9.3		Public relations and press relations	143
9.4		Advertising mobile services	144
9.5		Publicity	147
9.6		Sponsorship and product placement	148
	9.6.1	Viral marketing and communities	149
9.7		Conclusion	149

10 Branding   151

10.1		What is a brand?	152
10.2		Why brand?	152
	10.2.1	Brands aid in decision	153
	10.2.2	Brands and teenagers	154
	10.2.3	Brands and price	154
	10.2.4	Brand and loyalty	155
10.3		Needs to be comprehensive	155
	10.3.1	Brands in mobile telecoms	156
10.4		How to build a brand	157
	10.4.1	Where do I begin?	157
	10.4.2	Employee buy-in	158
	10.4.3	Damaging the brand	158

	10.5	Multiple brand messages	158
		10.5.1 Cross branding	159
		10.5.2 Sub-branding (overall company branding versus product trademarks)	159
		10.5.3 Co-branding	160
		10.5.4 On-line branding	161
	10.6	Action plan for branding	162
		10.6.1 Branding 'do's'	163
		10.6.2 Branding 'don'ts'	164
		10.6.3 Brand development plan outline	165
		10.6.4 Brands grow too	166
		10.6.5 After the brand, what is left?	166
11	**Service Adoption**		**167**
	11.1	S-curves	168
	11.2	Where is the saturation level?	169
		11.2.1 TV set analogy	170
		11.2.2 But can you use two phones at the same time?	171
		11.2.3 Subscriptions and subscribers	171
		11.2.4 So where is the ceiling?	172
		11.2.5 'Near saturation' myth	173
		11.2.6 An American consideration	173
		11.2.7 How high is high?	174
	11.3	Business or Residential	174
		11.3.1 The case for business customers	174
		11.3.2 The case for the residential customer	176
		11.3.3 Exceptional issues with 3G	177
	11.4	Early adopters	179
	11.5	Mass market	181
	11.6	The early eight	182
	11.7	Beyond the adoption	186
12	**Reachability**		**189**
	12.1	Wireless carriages and voice telegraphs	190
	12.2	Enter reachability	191
		12.2.1 Calling the person, not the place	192
		12.2.2 Change plans	193
		12.2.3 Indispensible	194
	12.3	Reachability and mobile services	194
		12.3.1 SMS text messages and reachability	195
		12.3.2 Respecting privacy	196
		12.3.3 Knowing who calls	197

12.4	Cellular is a distorted case of Metcalfe's law	197
	12.4.1 Hockey stick is not Metcalfe's law	198
	12.4.2 Inflection points for the hockey stick curve	200
12.5	Most personal device	200
12.6	Reach out and touch	201

13 Selling Mobile Services — 203

13.1	What do you sell in 3G?	204
13.2	Selling through distributors	204
13.3	Selling to consumers	206
	13.3.1 Event related sales	206
	13.3.2 Bundling services with the subscription	207
	13.3.3 Billing inserts	207
	13.3.4 Portal placement	207
	13.3.5 Selling to businesses	208
	13.3.6 Corporate customers	208
	13.3.7 Large corporate customers	209
	13.3.8 SME or medium sized companies	209
	13.3.9 SOHO or small businesses	210
13.4	Selling to partners	211
13.5	Motivating the sales representative	212
13.6	Handset subsidies	214
13.7	Non-traditional sales	216
	13.7.1 Cross-selling	216
	13.7.2 Bonus point programmes	217
	13.7.3 Network effect/viral selling	217
13.8	Sales out	218

14 Tariffing — 219

14.1	But isn't tariffing simple?	220
	14.1.1 Cost-plus tariffing	220
14.2	Some customers are willing to spend more	221
	14.2.1 Airline analogy	221
	14.2.2 Applying the example to telecoms	225
14.3	Profit and pricing	226
	14.3.1 Prices and usage	228
	14.3.2 The variety in acceptable price	229
	14.3.3 Prices for service introduction	230
	14.3.4 Penny for your thoughts	231
	14.3.5 Pricing of bundles ('service packages')	232
14.4	Preparedness for tariffing	232
	14.4.1 Marketing research	232
	14.4.2 Tariff modelling	233

	14.4.3 Tariff trials	234
	14.4.4 Tariff adaptation	234
14.5	How about one price for all?	234
	14.5.1 Pricing by data traffic	235
	14.5.2 Home zones and hot spots	236
14.6	3G licences and the price	237
	14.6.1 Price 3G for mass market adoption	237
	14.6.2 Not a free for all	238

15 Billing 239

15.1	Charging, billing, reporting	240
	15.1.1 Charging collects the data	240
	15.1.2 Billing creates the invoice	241
	15.1.3 Reporting gives information to the caller	241
15.2	Micropayments	241
	15.2.1 Credit risk	242
	15.2.2 To bank or not to bank?	243
	15.2.3 Tracking advertising and promotion revenues	243
	15.2.4 Tracking digital rights	244
	15.2.5 Billing can also be an added value service	245
15.3	From billing to product management and marketing	245
15.4	The call for one bill	246
15.5	Revenue assurance	247
	15.5.1 Revenue leakage and profit	248
	15.5.2 Revenue assurance and 3G	248
	15.5.3 Billing complaints	249
15.6	End to billing	250

16 Other Revenue Streams 251

16.1	Redefining the operator position	252
16.2	Business models	253
	16.2.1 Case Jippii Group	253
	16.2.2 Case Sonera Zed	255
	16.2.3 Case I-Mode	256
16.3	Operator revenue strategies	257
	16.3.1 Selling location data	259
	16.3.2 Location based push services	259
	16.3.3 m-Commerce	261
	16.3.4 mAd (mobile advertising)	261
16.4	Revenue sharing	263
	16.4.1 Revenue sharing levels	264
	16.4.2 More money?	266

17	**Combatting Churn**			**267**
	17.1	Basics of churn		268
		17.1.1	Who is a churner?	268
		17.1.2	Why customers churn – three general reasons	269
		17.1.3	The joiner	269
		17.1.4	The leaver	270
		17.1.5	The changer	271
		17.1.6	Selecting customers to target	271
		17.1.7	Stayers	272
	17.2	Churn is good — targeting competitors' customers		272
	17.3	Churn is bad — don't let valuable customers churn		273
	17.4	Combatting churn		274
		17.4.1	Price as the weapon	274
		17.4.2	Technical barriers and churn	275
	17.5	Number portability		276
		17.5.1	Changing numbers	276
	17.6	Loyalty programmes		277
	17.7	Handset subsidies		277
		17.7.1	What makes subsidies so damaging?	279
	17.8	From techniques of authentification to identity		280
	17.9	Back to the brand		281
	17.10	Community think		282
		17.10.1	Customer intelligence and churn	284
		17.10.2	Keeping customers happy	285
	17.11	An end to churn		286
18	**Marketing Plan**			**287**
	18.1	Business, marketing, advertising plans		288
		18.1.1	Business plans	288
		18.1.2	Marketing plans	288
		18.1.3	Hierarchical nature of plans	289
		18.1.4	Segment marketing plans	290
	18.2	Marketing plan outline		291
		18.2.1	Plan ahead	293
19	**Postscript**			**295**

Being Part of the 3G Revolution	**301**
Abbreviations	**303**
Bibliography	**307**
Useful Websites	**309**
Index	**313**

About the Authors

Tomi T Ahonen is an Independent 3G Consultant based in London. He has presented at 100 conferences in six continents and lectures at Oxford University's 3G courses. He previously headed Nokia's 3G Business Consultancy Department and oversaw Nokia's 3G Research Centre. He also worked as Nokia's Segmentation Manager. Earlier he worked for Elisa and Finnet International in Finland and OCSNY in the USA. Tomi's accomplishments include the world's first fixed-mobile service bundle, the world record for taking market share from the incumbent, and a multi-operator billing system. He started  his career on Wall Street. Tomi holds an MBA from St John's University NY, and a Bachelors in Marketing from Clarion University. His previous books are *m-Profits* and *Services for UMTS*.

Timo Kasper is the Director, Finance, Administration and ICT (CFO) of Observer Finland, a company specialising in Competitive Intelligence and Business Environment Scanning services. Timo has been responsible for constructing winning strategies for the emerging competitive markets in Finland and deployed innovative IT and telecoms solutions. Earlier he worked as the Business Controller for Computer 2000 (aka TechData), Finland's largest IT wholesaler, and started his career as an Auditor and Consultant at Arthur Andersen & Co. where he participated in a wide range of attest service assignments, mergers and acquisitions, legal and financial due diligence and business process re-engineering projects. Timo holds a B.Sc. (econ.) degree from the Helsinki School of Economics and Business Administration and lives with his family in Espoo Finland.

Sara Melkko is an Independent Expert in the field of telecommunications and the Internet, having worked in Finland and Germany as Marketing and Brand Manager with the digitally converging solutions provided by Elisa Communications across fixed, mobile and broadband telecoms, and then with pioneering mobile Internet innovations with Jippii Group. Sara then worked with IT security matters with SECUDE GmbH. Sara has a multicultural and -lingual background and holds a Business Administration degree from the University of Passau in Germany and has completed the "European Master's Degree in Human Rights and Democratisation" Programme in Venice and Padua, Italy. Sara currently focuses on Human Rights Business perspectives within Corporate Social Responsibility. She currently maintains home bases in Frankfurt and Helsinki.

Foreword

Calling a person and not a place, such as a fixed line in the home or office, is still not well understood as a concept, even if the adoption of mobile telephony now exceeds 1.5 Billion customers worldwide, or a quarter of the world's population. However, selling phones is quite different from selling services and it is in the latter area we have much still to learn.

As the Mobile industry approaches a $1000 Billion (or $1 Trillion) global industry, 3G will play a much bigger part, providing for much more Capacity (spectrum, coverage, and quality), Capability (higher-speed data access, personalisation, IP-based Applications) and Content (Entertainment, Business information and access to a huge Mobile library or "Web"). These 3 C's will not be sold (or accepted) as Technology, but as features or services that will require professionals able to understand how to do their Mobile Marketing.

The authors wrote this book to "help marketing managers bring modern marketing methods to mobile telecoms." In this respect they have succeeded in bringing together key subjects/lessons for the marketing of advanced mobile services/applications, grounded in strong experience of modern second-generation data services.

The introduction to "murfing" (or mobile surfing) of the web and "ta voitettavuus" (or reachability) are just two concepts introduced against an extensive menu of marketing techniques. These are all set against a backdrop of significant growth in usage of Mobile services globally and a need to professionalise usage segmentation, market intelligence, and partnerships.

The authors also bring out some of the reasons why mobile is different, personal, and independent of location. Techniques such as the "5 M's" of mobile service creation and the "Early 8" candidates for new service development are just two approaches that are offered to the reader to identify and implement leading data services for mobile markets.

In the same way that Transportation choice revolutionised the 20th Century, with investment in railways, automobiles, and airplanes, the Communications choice is set to have similar impact on the 21st Century. It is very important that wireless (with its roots in the late 19th century) and mobile, in particular, play their part. With technical innovation running faster than before, it is more important than ever to understand the role that mobile will play in the many Communication markets it can address. This book provides a strong introductory framework to the mobile marketing within this broader Communications revolution.

Mike Short
Vice President, Research and Development – O2
Chairman, Mobile Data Association
Past Chairman, GSM Association

A real friend is one who walks in when the rest of the world walks out.
—Walter Winchell

Acknowledgements

We Thank You

This book was a collaborative effort by the three of us but in the same way that 3G was a convergence, the ideas we discuss in this book reflect a combined know-how gathered from widely dispersed areas. We would like to give particular thanks to specific people involved with this project.

When we were researching this book there were no business books on 3G. The focus for the book took many turns and evolved along the way. We owe a debt of gratitude to Jouko Ahvenainen, the 3G segmentation guru of Xtract, who helped us plan this book. Several other people have had a profound impact on how this book appears, without always even knowing they had done so. In assistance to the concepts, structure and focus of this book, we want to especially thank Linda Charnley of Vodafone; Steve Jones of the 3G portal; Helena "Hekku" Kahanpää, Ebba Dåhli, Mikko Lavanti, Merja Vane-Tempest, Julian Heaton and Janne Laiho of Nokia; Matti Makkonen of Finnet; Taina Kalliokoski of TeliaSonera; Matti Tossavainen and Teppo Turkki of Elisa; Sharon Haran of Partner/Orange; Voytek Siewierski of NTT DoCoMo; David Leshem of Compwise, Marc Hronek of MetAccord; and Joe Barrett of Flarion. We hope you can discover that we have tried to incorporate your thoughts and suggestions.

We also want to thank several other experts who have provided us with valuable advice relating to the issues in this book. We want to thank Joe

Grunenwald of Clarion University; Helmut Schmalen of the University of Passau; Juha Malmberg, Juha Kervinen, Olli Rasia, Tiina Kovero and Katja Laine of Elisa as well as all of the old LLL gang known as "Lällärit"; Minna Rotko of Radiolinja; Jokke Viitanen of Song; Anne Nikula of Jippii; Seppo Hakanen Auria/Turun Puhelin, Markku Lempinen and Tarja Aarnio of Finnet International; Annele Aerikkala, Pirjo Kainu, Päivi Klinga and Tommi Vihavainen from Observer Finland; Harri Poimaa from Deloitte; Johan Wallin from Synocus; Roberto Saracco of TIM; Mark Weisleder of Bell Canada; Mike Short of O2; Tarja Sutton, Paavo Aro, Teija Hyttinen, Minna Sainio, Nicole Cham, Russell Anderson, Ilkka Pukkila, Merja Kaarre, Kati Holopainen, Paolo Puppoli, Rob Hughes, Tuula Putkinen, Arja Suominen, Lauri Kivinen of Nokia; Margit Brandl, Eva Remerie and Martin Moedl of Siemens; Sonja Berger of IMG; Jyri Loikkanen of Starcut; Todd Stevens of OCSNY; Steven Chan of M1; Paul May of Verista; Bengt Nordstrom of Northstream; Kimmo Kiviluoto and Jari Saramäki from Xtract; Tom Hume of Future Platforms; and Alan Moore and Axel Chaldecott of SMLXL. Also we want to thank some experts who are independent today: Risto Linturi, Lars Kerschbaum, Stephen McLelland, and Nick Frengle.

At our publisher, John Wiley & Sons, Ltd, we most want to thank Sarah Hinton for her patience and support throughout this lengthy project. Also our warmest thanks to Mark Hammond and Geoff Farrell, who have always been on our side. And a very special thank you to Peter Holland at Oxford University, who has supported us in so many ways.

Finally, this book has taken a huge toll on our private time. Our nearest and dearest have sacrificed and compromised to enable this project to remain on schedule. We are most grateful for the patience, support and love of Outi, Olli and Salla Kasper, Antero and Liisa Kasper, Uki and Ulla. While scattered in different countries, the support and communication via different media with your loved ones is a source of inspiration. We thank Tiina, Heikki, Jukka, Hanna, Jari, Inkeri, Tepa, Kari, Pirjo and of course Jan and Ulla. Hoping that 3G will make our daily communication easier, quicker and more instantaneous and affordable, a special thank you is expressed to Päivi and Heikki Melkko, Samu Melkko, Tommi Melkko and Tommi Makkonen.

We know this project took a lot out of our families and dearest but we did try to keep this project in its proper balance with our other commitments and our lives. For the reader we suggest this thought from Peter Lynch: "I don't know anyone who on his death bed wished he had spent more time at the office."

Acknowledgements

We welcome feedback for this book, please send it to 3gmarketing@tomiahonen.com

Tomi T Ahonen, London
Timo Kasper, Helsinki
Sara Melkko, Frankfurt

It's kind of fun to do the impossible.
— Walt Disney

1

Introduction

For much of the 1990s, mobile telecommunications was an interesting sideline in the overall business of telecoms. The mainstream telecoms revenues came from long-distance and international calls on fixed (wireline) networks, and the Internet was seen as the future with rapid growth potential. Mobile (also known as wireless or cellular) telecoms was seen as a niche market with only limited growth potential, and in the early 1990s nobody expected the number of mobile-phone subscriptions to exceed that of fixed subscriber lines. In fact, the technical standards assumed penetration rates of no more than 20% or so. The mobile operators were left much to their own devices. By the end of the 1990s, as the fixed Internet was starting to exhibit

---

3G Marketing: Communities and Strategic Partnerships Tomi T. Ahonen, Timo Kasper and Sara Melkko
© 2004 John Wiley & Sons, Ltd ISBN: 0-470-85100-7

signs of major problems in achieving profitability, the mobile telecoms business emerged as the healthiest and most robust of the telecoms sectors. Were it not for the huge licence fees paid for next-generation 3G licences in Europe, the mobile phone business may have even escaped the downturn that hit the rest of telecoms in 2001 and 2002.

1.1 A look back

The first cellular phone call (mobile-phone call) was placed in the USA in 1973 by Motorola's Martin Cooper. Despite having all the best elements for technological leadership in this area, such as a powerful IT industry, open telecoms competition, a wealthy and automobile-crazy population (the mobile phone was first used primarily as a car phone), and a culture for competition, the USA was never again the focal point for innovation in cellular telecoms. The first commercial cellular network was launched in 1979 in Japan by NTT, three years before the first cellular network went commercially live in the USA.

The first internationally roaming cellular network standard was NMT (Nordic Mobile Telecoms), introduced in 1981 throughout Scandinavia. The first digital cellular network, GSM (Global System for Mobile communications) was opened commercially in Finland in 1991 by Radiolinja, ushering in the age of competition in mobile telecoms. Many valuable lessons for the evolution to 3G can be gained from examining how the leading countries and companies experienced the changeover from 1G to 2G. It is

Figure 1.1 Users on mobile and fixed networks (UMTS Forum August 2003)

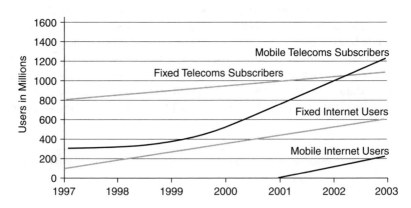

interesting to read how current critics evaluate the potential for 3G, in light of what experts said about 2G 10 years earlier. The network was patchy, of inferior quality to the analog systems of 1G, the handsets were more expensive, had availability and quality problems, etc. No valid services seemed to suggest that there was a solid reason to switch from 1G to 2G. While it was a dramatic change in the technology of cellular telecoms, digital networks by themselves did not introduce any new capabilities. The first digital services were digital versions of voice services, namely voice calls. It was not until in 1994 that the first new digital service, SMS (short message service) text messaging services were introduced in Finland and the UK.

Nobody expected SMS to become a major part of services used, or to deliver significant parts of operator revenue. Certainly nobody in the mid-1990s expected any billable added-value services and mobile commerce to be conducted with SMS. Most experts dismissed SMS as so inconvenient to use as to be totally unusable and unattractive to the general public. Many analysts argued that SMS would cannibalize voice traffic. The attraction of SMS was assumed to be limited only to the poorest segments of users, including the youth.

The success of SMS took the whole industry by surprise. Many GSM phone manufacturers did not provide SMS sending as a standard feature on all early phones, illustrating how little they believed in the service. Most operators did not build SMS gateways for enabling text message traffic into the early GSM networks. Not until the early news started to filter out of Finland and Sweden, that over half of the user base was getting addicted to SMS and that they were sending more than an SMS message per day, did other countries adopt SMS. After 8 years, by early 2002, SMS usage was measured in billions of messages sent. By the spring of 2003, SMS revenues accounted for about 15% of operator revenues across Europe, and as much as 35% of revenues in the leading country in Asia, the Philippines. For some segments, in particular youth, SMS had become the killer application why they wanted a mobile phone: they didn't want to call someone, they wanted the ability to send secretive messages to their close friends and connect with their buddies, their personal communities.

The mobile telecoms industry was slowly adding new services. In 1994 Nokia installed the first pre-installed (music) ringing tones to selected models, and soon such features of the phone handset emerged as interchangeable covers and pre-installed games. In 1996 the first prepaid services were launched by Telecom Italia Mobile (TIM) in Italy. This was key to enabling the 'whole population' of any Western country to get on mobile networks. Next, in 1998, the mobile telecoms industry faced another innovation, and the mobile data value-add services industry was born. This was invented by a surprising player in Finland. In the advanced technology hotbed and in an environment of lively telecoms competition, these services were *not* introduced by a mobile operator.

What is now called the mobile Internet or the wireless services market, and where the current ultimate pinnacle is 3G, was invented by a fixed Internet service provider (ISP). In 1998, Saunalahden Serveri (now called Jippii Group or just Saunalahti), one of the biggest ISPs in Finland, launched a gimmicky service to manipulate the software on selected high-end Nokia mobile phones. The service ran on top of SMS, and Saunalahden Serveri had negotiated an interconnect deal with the incumbent mobile operators in Finland to get part of the SMS revenue for these messages. The messages allowed users to send new codes for installing custom ringing tones on mobile phones. The first versions of the solution had a small selection of pre-programmed songs, and also allowed users to compose their own songs and use them as ringing tones. The *downloadable* ringing-tone business was invented. Under the noses of the world's leading mobile operators in the most advanced mobile telecoms country, a fixed ISP created the mobile services market.

The service idea spread like wildfire around the world, as everybody wanted the latest hits by the Spice Girls, Backstreet Boys, Madonna and Eminem, etc., on their phones. In three years the ringing tone market had grown from zero to over 1 billion dollars worldwide and by 2003 at 3.5 billion dollars ringing tones generated 10% of the total revenues of the music industry. As the first value-add downloadable service, ringing tones were also ahead of the curve among all mobile content and services; Still in 2003 ringing tones represented the most popular or second most popular added-value service on all of the world's mobile networks. Very closely upon the heels of the ringing tone market, came the downloadable logo market, also pioneered by Saunalahden Serveri.

With SMS emerging as a viable delivery vehicle for billable added value services, the Scandinavian operators competed with innovations of all that could be delivered on SMS. Solutions on car parking were pioneered in Norway; paying for vending machines to buy Coca Cola and paying for car-washes were introduced in Finland; running national polls and surveys on SMS appeared in Sweden. The text-based simple mobile service market was exceeding all expectations.

Introduction

1.1.1 Enter I-Mode

In 1999 the cellular innovation pendulum swung suddenly back to Japan when NTT DoCoMo launched its revolutionary I-Mode service. I-Mode and its rivals from the two other leading cellular operators KDDI and J-Phone (renamed Vodafone KK) in Japan, quickly introduced ever smaller phones, colour screens, polyphonic sounds, MIDI interfaces, built-in cameras, and numerous examples of games, mobile commerce, information services, and countless entertainment services.

A major part of I-Mode's success was the way it attracted content partners. Perhaps due to the Japanese business culture of working in collaborative communities, NTT DoCoMo set up a partnership structure that was very radical compared with any other contemporary thinking in telecoms. While most telecoms operators around the world were offering revenue sharing models for content revenues of 50/50, or even 75% for the operator and 25% to the content owner, in I-Mode, NTT DoCoMo set its fee at only 9%, with 91% of the content revenues going to the content owners. (It should be noted that almost no operators in mobile telecoms ever share in transport or airtime revenues, which may be well in excess of the content cost for transmitting data on a mobile network).

NTT DoCoMo's 9% cut of content revenues was seen as a fair and non-intrusive fee to pay to get content onto its I-Mode service. Due to this approach, the amount of content providers has grown dramatically. The content providers and application developers in Japan have created a broad array of services ranging from the useful — such as translation services — to the frivolous, such as the fishing game. With more content, users have found even more reason to sign up and use the service, and, very importantly for success in mobile services, to spread the word. NTT DoCoMo had the largest number of mobile subscribers in Japan, yet another reason for more content providers to join. And with ever more content, more subscribers: a virtuous cycle.

While the revolutionary revenue-sharing scheme is not the only reason for I-Mode's success, most analysts rate it as one of the most important, if not the most important. The service has also benefited from being based on HTML, on the fact that the fixed internet penetration in Japan was very low, that NTT was the market leader in Japan, that SMS was not a viable rival, etc. What I-Mode did, more than anything else, was to validate the fact that the 'Oyayubizoku' or the 'Thumb Tribe' is viable — that mobile phone users can take the phone from the ear and also start to consume services from colour screens. The Thumb Tribe also was one of the first signs of the new communities that were emerging connected only by their mobile phones.

1.1.2 The flap about WAP being a failure

During 2000, in quick succession, most advanced countries introduced WAP (Wireless Application Protocol) services. As WAP was strongly hyped to bring the Internet to the pocket, there were very high expectations. With disappointing tiny screens in monochrome, slow connections and almost criminally high usage prices, WAP failed to ignite the public's affection. In European countries the mobile operators were very selfish in their revenue-sharing propositions, and with unfavourable terms were unable to attract vast ranges of new content. Where WAP generally is criticized as a failure, the technology itself is not at fault. In Japan, J-Phone and KDDI have used a local WAP variant to considerable success, and Korea has had similar experiences. The biggest single key in why WAP was seen as a failure in Europe is most probably the pricing, with users having to pay for all the waiting time, thus making the first services seem prohibitively expensive. The bad word-of-mouth spread, and few users in Europe warmed to WAP. Interestingly, in advanced Asian countries and to a lesser extent also in the USA and Canada, WAP was seen as more of a success.

During 2001 GPRS services, classified soon as '2.5G' — as an interim upgrade technology to 2G before the replacement cellular technology of 3G — emerged and soon after came also CDMA 2000 1×RTT, another 2.5G technical standard. These technologies allowed faster speeds for data use, and the introduction of 'packet based' datacoms services similar to those with computers and the Internet. In close harmony with 2.5G technology, various advances were made in other mobile phone related technologies, such as the introduction of colour screens, polyphonic ringing tones, integrated digital cameras, etc.

Then, in October 2001, Japan further advanced its lead in advanced cellular services with NTT DoCoMo introducing its 3G services, branded FOMA. 3G services on CDMA 2000 EV-DO standards soon followed in Korea, and 3G was launched through various 'soft launches' and 'friendly user trials' throughout many leading European countries in the Autumn of 2002. 3G arrived commercially in Europe in March of 2003, when Hutchison's service branded '3' was introduced in Italy and the UK. By the end of the year 2003 there were 5 million subscribers to 3G, mostly in Japan and Korea, but also about half a million in Europe. The latest thinking as

Introduction

the book was going into print was that apart from the famous three Gs of 3G — Games, Gambling and Girls (adult entertainment) — the next big thing would be picture messaging and other multimedia messaging or MMS, but only time will tell.

1.1.3 Growth rate

It is typical of the mobile telecoms industry that innovation with advanced mobile services has appeared suddenly and often in surprising regions (Table 1.1). The growth of mobile telecoms during the 1990s, especially in Scandinavia, Italy, Hong Kong and Taiwan surprised everybody, including of the operators in those leading countries. The hyper-growth of the late 1990s had mobile operators throughout Europe and Asia rushing to connect subscribers, with little effort being needed in marketing. Hence the industry grew past the revenues of the fixed Internet and soon rivalled the revenues of fixed telecoms, without putting much thought into modern strategic marketing tools and methods.

As subscription penetration has reached more than 90% in several European and Asian countries, and increasingly the 20–35-year-old population has two, sometimes three, mobile phones, the industry is reaching a more mature stage. The times of hyper-growth are over, and now most new subscribers for one network are no longer first-time mobile phone users,

Table 1.1 2002 Penetration rates of largest mobile countries

Country	Subscribers (million)	Penetration (%)
China	190	15
USA	137	48
Japan	78	61
Germany	57	69
Italy	53	92
UK	49	83
France	38	62
Spain	32	81
Korea	32	67
Brazil	32	18
Mexico	25	25
Taiwan	24	105

Source: Merrill Lynch Wireless Matrix December 2002

but rather users who have switched from another network. Churn has entered the mobile telecoms market place. Still, with dozens of countries reporting penetration rates of over 90% and Taiwan, Israel and Italy at over 100%, the industry still has plenty of growth. While subsriptions may not grow dramatically, the mobile usage, traffic and revenues through more advanced services have plenty of opportunity to grow during this decade.

1.2 What have we learned?

The cellular phone is over 30 years old. The mass market for mobile phones is about 12 years old. SMS is 10 years old, and the mobile Internet or advanced mobile services business is only 5 years old. As sociologists tell us, it takes anywhere from 3–6 years to cause societal change of behaviour, and we have only recently started to see how much mass-market voice services on mobile phones have changed our habits and culture. Some of the early arguments centred on whether mobile phones can be used in trains and other public transportation, for example, with, of course, opposite decisions having been taken on the matter such as in Finland where society adjusted and accepted the practise, as on a typical rush hour bus you might hear five simultaneous ongoing conversations each in a subdued voice, while in Japan nobody talks on the phone in their trains. In both countries more people use text messages and various mobile services for entertainment, communication and work.

SMS effects are so recent that only the very earliest adopter countries like Norway, Sweden, Finland and Italy can start to have meaningful data to analyse. Some very interesting academic studies are emerging on the sociological and cultural changes that text messaging have caused, such as Timo Kopomaa's findings which he published in his book, *The City in the Pocket: the Birth of the Information Society*. Also, the mobile Internet is much too young to have had a chance to prove itself. All stories about what the real impact of the mobile Internet supposedly is, are based on samples of very small size or only very specific segments of overall society. Yes, we see a lot of character downloads on I-Mode in Japan, and the traffic loads, customer satisfaction, etc., are obvious, but the data is based on young Japanese enjoying the novelty of consuming Disney on their current favourite gadget, the mobile phone. Only when we see how well the full population adopts and starts to use I-Mode content can we start to make assumptions about similar societal impacts in other countries. Will grandmothers also find compelling services and content from advanced wireless services?

A few early successes have emerged, but it will take a lot of trial-and-error to capitalize on the opportunities of advanced mobile services. This book includes a chapter on mobile service creation, but for those who are serious about how to create mobile services, we recommend the book *Services for UMTS* by Tomi T. Ahonen and Joe Barrett. For those who need to evaluate the traffic loads of new services to next generation cellular networks, we recommend the book *Radio Network Planning and Optimization for UMTS* by Laiho, Wacker and Novosad. A significant key to attracting the best content and application partners is an attractive revenue-sharing plan and pricing below the pain threshold. For readers who want to understand the revenue-sharing, pricing and other money issues of new mobile services, we recommend the book *m-Profits*, by Tomi T. Ahonen. For those who need to understand the UMTS system, we recommend the 'bible' of the industry, *WCDMA for UMTS* by Harri Holma and Antti Toskala.

1.2.1 Telecoms operators and 3G marketing

The telecoms operators (wireless carriers) were caught in a phenomenal rate of growth in the industry throughout the 1990s as the mobile phone went from being an exclusive and expensive executive gadget to a mass-market, everyday, communication tool for everybody. During that rapid growth phase, mobile operators were hard-pressed to keep up with the unanticipated growth and struggled to deliver newer mobile phone handsets and connect customers, while expanding their network coverages. The mobile operators achieved remarkable market success with minimal modern marketing efforts. The pie kept growing so fast that all players kept receiving larger slices of the pie without having to resort to capturing customers from competitors.

Now during the first decade of the Twenty-first century, mobile telecoms is facing slower rates of growth, more competitors within the industry and rival technologies outside it. Customers are more astute, tend already to have their mobile phones and subscriptions, and now need to be convinced to *upgrade* rather than acquire their first mobile phones.

The mobile operators are discovering the harsh truths of the real competitive marketplace. As American business schools teach, 'Customers vote with their dollars' and those competitors who offer the best value proposition

Table 1.2 Simultaneous paradigm shifts from 2G to 2.5G/3G

From	To
Technology focus	Customer focus
Acquiring new customers	Reducing churn
Selling handsets	Selling services
Treating all equally	Targeted marketing
Market share in subscribers	Market share in traffic
Minutes of use (MOU)	Lifetime value
Do it yourself	Do it with partners
Revenue	Profit

to customers will gain, while the others will lose. The performance is relative to other players in the market, and the nature of competition is such that it never gets 'easier', but rather the competitive pressures bring ever more intense marketing efforts by all players involved.

The senior management heading marketing, product management, sales, etc., at mobile operators has not had to learn modern marketing methods to achieve the successes of the 1990s. They are therefore ill equipped to move their organizations to become successful in the rapid pace, customer and services-oriented market place of mobile telecoms in this decade. This book is intended to provide the basic guidelines for doing just that: it covers all the major areas of current theory of marketing and applies them to mobile telecoms. The authors of this book have been personally involved in deploying pioneering efforts in these marketing areas within the world's leading mobile telecoms markets.

1.3 Lets touch upon definitions of 3G

This book examines the best lessons from around the world and shows how advanced mobile services can be marketed. While we focus on 3G in this book, practically all of the concepts, theories and tactics can be used with so-called '2.5G' technologies and the upcoming 3.5G and 4G as well as many complementary/competing technical solutions such as W-LAN (Wireless Local Area Network) also known as WiFi (Wireless Fidelity) or by its technical standard 802.11; WiMax (802.16), Bluetooth, etc. In fact most of the theories and examples in this book can be applied to all telecoms marketing.

For a quick set of definitions, when we talk about 3G in this book, we include the standards W-CDMA and CDMA2000 EV-DO, CDMA2000

Introduction

EV-DV, and TD-SCDMA. Depending somewhat on the writer, some of the above are also called IMT-2000 and/or UMTS. When we talk about 2.5G, we mean packet-based digital networks that are compatible with the first digital cellular systems, primarily GPRS, EDGE and CDMA2000 1×RTT. For us, 2G means the first digital standards such as GSM, CDMA, TDMA, and PDC, and their basic data services such as HSCSD, WAP, SMS, I-MODE, etc. At the time of writing this book, the definitions of 3.5G and 4G had not yet been finalized.

The collective term 'Mobile Internet' or advanced wireless services or VAS (Value Added Services) in mobile telecoms or even the term m-Commerce (Mobile Commerce) can be assumed to include all of the above, with the additions of W-LAN, i.e. Wi-Fi, i.e. 802.11; WiMax (802.16); 802.20; and Bluetooth and IR (Infrared) enabled services and applications. The issues covered in this book can be used successfully in the marketing of any mobile internet service or proposition.

Finally, 1G covers the old analog cellular systems such as AMPS and NMT. The analog systems are being phased out, and 2 years ago Finland was the first country totally to turn off its 1G network.

Figure 1.2

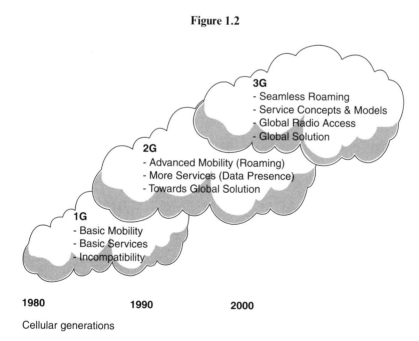

Cellular generations

As this book will not spend much time on the technical standards and any differences between them, there is no real reason to waste the reader's time in explaining what each of the above definitions means. For those who want to know, they are of course all included in the abbreviations section at the back of the book.

1.3.1 So what is 4G

As for 4G, a good deal of early argument has emerged as to what it is. Some self-serving statements have been introduced saying that some companies have 4G today and that 3G is obsolete because of it. Without going into technical details, such claims are very misleading. Let us illustrate via an analogy. If we talk about a new generation of air travel and promise to introduce a radical new type of airplane, and then we offer bullet trains to the major airlines of the world, then yes this is a new form of *transportation*, but it is not a new type of *air travel*. It will not replace the airplane. Bullet trains are excellent at something *different* to traditional airplanes, but a bullet train will not traverse the oceans. While arguably competitive in transport over short and even medium distances, the bullet train is not viable for intercontinental travel. Thus a bullet train is a partial replacement. It is not the same kind of travel. No airline would abandon its fleet of jet planes and switch only to bullet trains, but many smart airlines might find a good business opportunity to incorporate bullet trains for some of their busiest short-haul routes.

1.3.2 W-LAN or Wi-Fi is definitely not 4G

The analogy here illustrates that for us to *replace* the 3G telecoms *network*, we will have to provide a technology that is, in general terms at least, capable of replacing the previous cellular telecoms network. When some analysts or companies suggest for example that W-LAN (Wi-Fi or 802.11) based data networks would be 4G, this is vastly misleading and serves only to muddle the issue. While both are wireless, a cellular system is capable of handling dozens of millions of users placing telecoms voice calls while mobile from anything ranging from walking at the office to sitting in a speeding car. A system based on W-LAN is capable of handling high speed data at short ranges, but so far nobody has suggested that feasible W-LAN systems could handle voice traffic for example from moving cars on highways.

Thus any W-LAN system can be a partial replacement, and a complement, but it is not a total replacement to 3G (or 2.5G or 2G). Therefore W-LAN cannot be 4G. The easy confusion arises when the argument is presented solely on the side of datacoms speed. Yes, a W-LAN system can deliver much faster wireless data speeds than 3G, which in turn delivered faster wireless data speeds than 2G, but this was not the defining factor determining the first, second and third generations of mobile telecoms. Remember that the primary use of a mobile phone is voice, and it is likely to be so even toward the end of this decade. The 1G, 2G and 3G descriptions reflect a replacement of the cellular network infrastructure and the related handsets. A 1G mobile phone handset, such as an NMT or AMPS handset, cannot be used on a 2G network such as GSM, CDMA or TDMA. That is what the generations mean in 1G, 2G and 3G.

1.3.3 4G will arrive ten years from now

4G will be coming, of course. Just like 2G and 3G, 4G will require global telecoms standardization to approve the technical specifications and allocate the radio spectrum. The current understanding is that the spectrum allocation will happen at a World Radio Conference in 2006 or 2009, with the initial standard for 'Systems beyond IMT2000' being developed during the following few years to be ratified by the ITU (International Telecommunications Union) which is the topmost governing body to ratify all telecom standardization. The earliest that the first prototype 4G cellular networks can be deployed would be around 2010, but more likely 2012. If anybody suggests that they currently have 4G prototypes in use or being developed, are just capitalizing on technology hype without being in touch with cellular telecoms reality.

1.4 To sum up

This book is about marketing new mobile services. It assumes that the radio network is functional, the services are created, the content contracts are signed and the terminals are available. When the preparatory work is done, the sales effort will start and the real 'war' for customers and traffic can start. This book is intended as a handbook to guide in that contest. Tracing the various marketing activities from understanding customers to raising awareness to closing the sale and keeping customers satisfied, this is the book on the 'how' of wireless services. With ever more enlightened customers becoming increasingly selective with mobile operators, and as more players enter the mobile telecoms markets, the marketing tools and processes need to be increasingly refined and upgraded. Operators must learn to stop trying to offer everything to everybody under identical terms. The real world of profitability and customer orientation requires focused marketing efforts. Focusing cannot be done without segmentation. We could summarize this whole book into that one word, segmentation. It is the key to satisfying customers and to optimal tariffing. We will discuss segmentation in its own chapter and return to that concept throughout the rest of the book.

The advanced mobile services market is estimated to reach 1 trillion dollars (1000 billion) by 2010. As the advanced mobile services industry was invented only in 1998, in order to reach such numbers the industry will experience greater sustained growth than ever seen before, quite literally combining in the mobile Internet the speed of mobile telecoms with that of the fixed Internet. To achieve success, attract customers, generate revenues and make profits, it will be necessary to be swift and move boldly. A lot of experimentation is needed and success is not possible without learning to partner successfully.

The new mobile services industry will be fighting against the entrenched 'conventional wisdom' in telecoms, the Internet, and the content industries. Many innovators will face obstruction from those who say it cannot be done, it is a silly idea, there is no market, etc. Others will fear the mobile services as threats to cannibalize current business methods, products and services. Many will argue against the mobile phone as being a fashion item. Others will insist that there is an arbitrary ceiling to mobile phone penetrations. Still others will claim technical superiority of other solutions. Because the speed of growth will be so great, winning in this environment requires the courage to take a bold stand and follow-through. For that, we remind the reader of what Dale Carnegie said, 'Most of the important things in the world have been accomplished by people who have kept on trying when there seemed to be no hope at all.'

Your most unhappy customers are your greatest source of learning.
— Bill Gates

2

Market Intelligence
Before You can Act, You Must Understand

Resilient, flexible and intelligent organizations will be the leaders of future business. The leading companies will possess tools, processes, employees and organizations designed for the cutting edge of competitiveness. The ability to react to changing businesss conditions and to adjust to meet new market opportunities is pointless if the company does not possess a good understanding of that environment. Much as a nimble racing car is

---

3G Marketing: Communities and Strategic Partnerships   Tomi T. Ahonen, Timo Kasper and Sara Melkko
© 2004 John Wiley & Sons, Ltd   ISBN: 0-470-85100-7

worthless if the driver doesn't know the race track, so too must an analysis of marketing start with market intelligence.

2.1 What is market intelligence

We often hear people in marketing use terms such as market research, competitor analysis, market intelligence, competitive analysis, business intelligence, etc., as if they were synonyms. While all of them relate to understanding the elements that a business is in, it is important to understand how they differ. It is similar to understanding how business disciplines like accounting, marketing and management differ from each other.

2.1.1 Evolution of market intelligence

Management theory has been promoting different 'business drivers' during the last 60 years (Figure 2.1).

Back in the 1940s, mechanical technology was considered to be one of the most crucial factors in competitiveness and business efficiency. Investments were the focal point in the 1950s and 1960s. 'What happened in the 1970s'? you ask! No, the hippies didn't steal it; yes the Beatles broke up, but the energy crisis and recession did not allow any specific theory to dominate the general discussion.

2.1.2 Information, analysis, knowledge and intelligence

Analysis, knowledge and intelligence are often used interchangeably. For example, some companies have units devoted to competitor information, others call them competitor analysis, others identify similar units as competitor knowledge, while yet others call it competitor intelligence. These four

Figure 2.1 Business drivers promoted by management theory

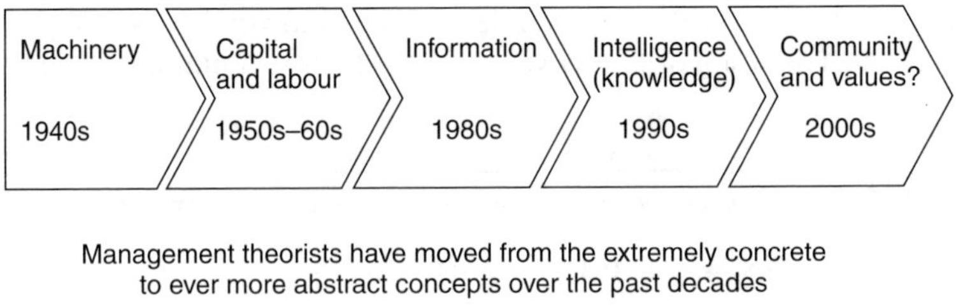

Machinery	Capital and labour	Information	Intelligence (knowledge)	Community and values?
1940s	1950s–60s	1980s	1990s	2000s

Management theorists have moved from the extremely concrete to ever more abstract concepts over the past decades terms are not exactly synonyms. Information is a collection of data with no analysis or understanding. Analysis goes a bit deeper, providing thought and understanding of the information but still provides no deep insight. Knowledge requires an understanding of the relevance of the information. Intelligence goes even further and requires providing the business impact of the information. The four terms are progressively more insightful, and require even more understanding of both the information and its relevance to the business.

Let us use an example to illustrate the differences. If we find that a competitor has 500 000 subscribers, that is competitor *information*. If we note that the subscriber amount grew by 100 000 from 400 000, the information turns into competitor *analysis*. If we provide the perspective that this competitor grew by 100 000 while the other big competitor grew only by 20 000 in the same period, then the information has turned to competitor *knowledge*. If we know that of the 100 000 customers, 25 000 came from our company as our churn, and after we examine the underlying reasons for this kind of reaction in the clientele, derive the strategic cornerstones for that particular competitor, then and only then has information turned into competitor *intelligence*. This substance of intelligence is our company's stepping stone to improved competitiveness and it helps us to make the right strategic decisions.

2.1.3 Knowledge or Intelligence

Knowledge is *not* another term for intelligence. Intelligence is, in this context and in broad interpretation:

> The ability to change one's behaviour based on past experience and variate response to changing environment.

Knowledge is something that is built into intelligence, but does not necessarily insist on acting upon itself. Knowledge is more passive than intelligence, which indeed leads into action.

2.2 Systematic market intelligence

Market intelligence is a series of processes rather than a function or drill. In most companies these processes are not distinguished nor are they systematically developed. Market intelligence is typically something that 'happens' to the company and is only as efficient as the key individuals involved in it.

2.2.1 Market intelligence and business intelligence

Market intelligence is an integral part of business intelligence.[†] Business intelligence is a larger concept covering other aspects of the business. Market intelligence processes address typically the following issues:

Market intelligence process	
• The legal and regulatory system	• Competitors
• Customers	• Resource markets
• Technical environment	• Reference market studies
• 'Environment scanning'	• Partnership intelligence/networking

2.2.2 Legal and regulatory intelligence

The legal and regulatory system affects the foundations of modern mobile service businesses. The deregulation of telecommunications business has made this all possible within a very short time. However, there are still many pitfalls to avoid as the legislation and legal systems are not keeping up with the technological development. For example, the copyright issues over material on the fixed Internet are very difficult to determine. Similar and more complex questions of rights and obligations within the wireless community will arise. The operators and other actors in the mobiles business will be facing, among other things, questions of individual privacy and confidentiality.

New technologies introduce new opportunities for creative businesses but these can also backfire. In Finland the adult entertainment industry is legal and provides a wide variety of adult-oriented services. In the 1990s the 'sex lines' with premium-rate phone numbers were well established. With the advent of SMS (short message system more commonly known as text messaging or texting) in the mid 1990s, the sex-line operators noticed that they could collect the GSM numbers of those customers who called from mobile phones, and send advertisements back to the mobile phones automatically, via SMS. The advertisers also got very creative with the messages and often they were sent as seductive love notes from the secret partner. This then proceeded to cause embarrasment and a lot of married men had to explain why they got so many explicitly 'intimate' messages signed by unknown females. This practice was banned by the consumer rights bodies — officials and operators were assigned the task of monitoring

---

[†]The concept of 'business intelligence' is very often discussed under 'competitive intelligence', therefore it is reasonable to consider these two terms as synonyms.

the compliance with the new regulations. What was good business practise for a niche market — the perfectly legal sex industry in Finland — was very bad publicity for the much larger telecoms industry.

We could say that the operators prudently became moderators preventing wireless-spam phenomena. (*'spam'* is the flooding of the Internet with many copies of the same message, in an attempt to force the message on people who would not otherwise choose to receive it. Most spam is commercial advertising, often for dubious products, get-rich-quick schemes, or quasi-legal services. 'Junk mail' is an often used synonym for spam.)

Another example of the regulatory environment and its crucial role may be found in some of the less developed countries. There are former Soviet Union member states that do not have a fixed network infrastructure in place. Therefore, the extremely expensive cabling and/or satellite linking for the basic network still needs to be financed and built. With normal competition it would probably not be possible to build a mobile network that could reach the critical mass necessary, i.e. achieve the break-even point for a business to be profitable. A governmental or foreign financing is always one option, but so is a monopolistic situation — another intervention tool for local authorities. We have seen a fierce battle against a monopoly that was allowed mainly in order to collect funds and finance the building of a modern nationwide telecommunication network. The question became highly political and raised patriotism to new heights as over 50% of the stock of the monopoly company was subsequently sold to a foreign telecoms syndicate.

The political and regulatory environment in these two examples dramatically affected the business of the operators. Let us assume that in the first case operator A — a challenger — had no profound experience in consumer services, and limited resources in the legal department. Operator A's awareness of the privacy and confidentiality issue was therefore limited. The sudden and unforeseen change in the legal structure caused them great loss of clients, as they could not guarantee and protect the privacy of their clients. At the same time, operator B with a long history in fixed networks and thus extensive experience within the consumer service business, announced that they could immediately block the subscriber number, so that the caller number wasn't shown to the addressee.

2.2.3 Customer intelligence

Customer intelligence is the process of building a deep understanding of the customers of a business. Customer intelligence can be built individually if appropriate and practical, or by grouping customers into 'segments' and 'clusters' where customers in one segment or cluster share similar attributes and are different from customers that are grouped into other segments or clusters. Customer intelligence is discussed in detail in Chapter 3 on Segmentation.

2.2.4 Competitor intelligence

Competitor intelligence is the process of tracking news and information about competitors. The primary objective of the competitor intelligence process is to reveal the strengths, weaknesses, opportunities and threats of each individual *competitor* — respective strategic cornerstones and current activities. Knowing your competitor(s) well makes it possible to identify your relative strengths and weaknesses, and to discover your competitive edges. Some basic understanding of competitors' strategies can sometimes be obtained from their mission statements, slogans and the internal values they cherish. For any player involved in the mobile Internet, the range of competitors is considerable and they will need to be grouped and prioritized. In a broad sense, for example for a mobile network operator with a 3G licence, competitors would include all other mobile operators with a 3G licence, mobile network operators with a 2G/2.5G or other licence, mobile virtual network operators (MVNOs), fixed Internet operators, fixed telecoms network operators, mobile and fixed portals, W-ASPs (wireless application service providers), content aggregators, mobile Internet content providers, mobile services IT integrators, and mobile equipment manufacturers/vendors.

Competitor intelligence processes and the methods used to gather relevant information are very varied. Common tools are of course annual reports, industry analyst comments, media and other publicly available documents. The single most used competitor analysis tool is the Internet, starting with the competitor's own pages. The common denominator with all of these could be 'public domain'. At the opposite end of the scale we find methods that border on industrial espionage and similar illegal and unethical issues. It is important to acknowledge that some state secret-service organizations are today providing intelligence services (spying) for their domestic industry leaders — as the French equivalent of the CIA (Direction Générale de la Sécurité Extérieure *www.dgse.org*) — has done.

As discussed before, competitor intelligence should be an ongoing process. As with any process, the quality of 'input' determines much of the quality of the 'output'. When dealing with information, the 'output' quality can sometimes be improved by increasing the quantity of 'input' information, but nothing can beat the value of accurate and precise information. Determining the usefullness of information before 'processing' it, can be simplified as below.

We have two types of information sources, *primary* and *secondary*, and in addition we can usually put the information on a timescale; in other words the information may refer to historical event(s) or to future plans and anticipations. Depending on the issue the balance of these factors should be evaluated.

Primary source means that the information is presented as *facts*, i.e. it is not filtered or otherwise altered and it comes from the originating source. In this category we have live speeches, personal observations and public reports such as annual reports (although some may beg to differ after the Enron case hit the news).

Secondary information may or may not (we don't know) be altered by opinion or otherwise, or it may be selectively extracted from a more comprehensive set of information. Such sources include newspapers, analyst's industry reports, TV documentaries and other recorded/edited TV and radio broadcasts.

The competitor intelligence process is an intensive and focused set of activities, producing valuable information about competitors and their plans and current activities. All this is done to support internal decision making, to make better decisions and to gain competitive advantage. The final aim is of course to bring prosperity to all stakeholders.

The 3G and mobile Internet environment is the ultimate convergence of voice and data, fixed and mobile, telecoms and Internet, and content and delivery. The emerging competition and partially competing-partially substituting services create a massive range of prospective competitor companies, services, offerings and solutions. While no significant competitors should be overlooked, the competitor intelligence process is far too complex and expensive to allow monitoring of all possible competitiors. For this particular reason, a simplified overview called 'environmental scanning' should be put in place. Environmental scanning is discussed briefly later.

2.2.5 Technical environment intelligence

Technical evolution, and especially the accelerating speed of it, has been a hot topic for ages. Most of the prophesies about the future have been focusing

Figure 2.2 Average retail revenue per minute in the UK (OFTEL August 2003)

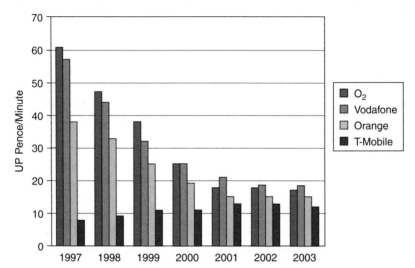

on quite 'big things', like colonies on planet Mars, daily space travelling for tourists, flying cars and underwater cities, etc. However, the greatest achievements have been in the 'small things', like DNA, molecular biology, computer chips etc. These 'little things' have changed our lives just as dramatically as the invention of number zero, another quite small thing.

In the telecom industry, the technological research and development has been so furious that very few good inventions have survived to be broadly adopted standards. Becoming an 'industry standard', is a key issue in building profitable businesses. How does this all relate to technological intelligence? The underlying question is: How can we identify the most prominent and suitable technologies for our business and how does that affect our strategy — in short term and in long term?

2.2.6 Telecoms is not used to rapid innovation

This is new for mobile telecoms. Up to the 1990s, telecommunications was entrenched in its standards-oriented evolution. As every telephone could connect to every other one on the globe, the telecoms network was — and still is of course — the largest interconnected device man has built. Any new telephone unit, public switching centre or other device had to conform to rigid standards in order to ensure that the whole system did not collapse. The standardization work became ever more complex and lengthy, and the 3G standardization work itself took about a decade.

As long as the only devices that man plugged into the telecoms network were telephones, and the only service provided was that of voice, and no new players entered into the field (as most markets were served by government monopolies), such a standards-based technological evolution could be maintained. Every new idea was put to committee and only after consensus with the other players would the new idea be accepted into the standard. Then every device had to be thoroughly tested before it could be installed. As non-voice devices were invented, such as the fax machine and computer modems, they too were strictly tested and their standards were as rigorously set as any other telecoms devices. Still, the modems were the first small ways that computers and the telecoms network started to work together.

2.2.7 The computer industry thrives on rapid innovation

The PC revolution was driven by very cowboy-like individual rogue ideas, through rapidly written codes, hastily constructed devices, with little if any testing, where users were expected to report bugs. This worked fine with inexpensive, often even personal, computers that were built by the end-users themselves. The philosophy that anybody could try to make anything spawned a huge industry, with companies literally in garages. The PC remained away from telecoms and on the desktop during the most of the 1980s, but in the 1990s, various networking technologies emerged. Most early networks were peer-to-peer and LAN (local area networks) networks where cables connected two or more PCs to each other. As the benefits of networking became obvious, so too became the understanding that larger networks yield greater benefits. Robert Metcalfe — who founded the networking company 3Com — formulated his famous law that the utility of a network grows as the square of the number of users. Thus a bigger interconnected network delivers more utility to each member than do smaller networks.

In the early to mid-1990s LANs were being connected to each other to form WANs (wide area networks). Where LANs were built with new wiring installed only to connect each PC to the other, and did not connect to the telecoms grid, the WANs connected via the telecoms network, and the then high speed connections such as ISDN. The PC had invaded the telecoms network. The real breakthrough was the emergence of the Internet as a global datacoms network. The Internet had remained a mostly university and research tool through the 1980s and early 1990s, until the web browsers emerged. In 1994 the growth was so fast that *Time* ran the internet as its cover issue. The internet brought about its protocol IP (Internet protocol) as the de facto datacoms standard. Now, with 3G services being based on IP, the merger of Internet datacoms and traditional telecoms is complete.

While the PC and Internet mindset provides for a lot of innovation and experimentation, it is frightening to traditional telecoms. Where previously there had to be a standard before any innovation, and because of the long lead times, vendors announced their technological road maps up to 3 years into the future, it was easy for telecoms operators to make technology plans 18 months ahead and longer. No surprises could emerge, as there was no way for any non-standard technology to be deployed, but IP is now a standard, and it allows any Internet services on any PC potentially to enter the arena. This creates a new headache for technological directors at the 3G operators.

Looking at some of today's technological 'hot topics', it is easier to understand the dilemma. Several companies are looking at XML-based document management, as that is supposed to provide another well-defined 'standard' for electronic communication, mainly by attaching a lot of information about the information itself in the message. Another hot topic is the programming platform — J2EE (Java 2 Platform, Enterprise edition). Should the user interface (or a thin client) be programmed to allow a wider variety of terminal devices (running different operating systems) to use the business logics in the application? To put it in plain English, the decision as to whether or not to buy a new DVD-video player instead of an old-fashioned VHS-VCR may constitute a technology intelligence problem. Remember what happened to SONY's better quality Betamax VCR system in the early 1980s.

The core of technology intelligence is, of course, keeping oneself up-to-date with the technical innovations and new applications within the industry. New technologies themselves, and sometimes even new applications of an old technology, bear higher inherent risk compared with well proven and 'standardized' technologies. However, it is clear that the leading edge companies seem to have very good relations with academic society pursuing 'pure' and applied science. The technology businesses support and finance a great deal of academic study — which finally should be of some benefit to the companies themselves.

Very plain SWOT-analysis (strengths, weaknesses, opportunities, threats) usually gives us adequate insight into emerging technologies. As mentioned above, in a fierce competitive environment companies regularly browse public domain information from patent registration offices and often also from university publications — such as master theses and other research. Increasingly the most relevant business-oriented research is classified by the universities as confidential for corporate sponsorship and other such reasons, and thus are no longer 'public domain'.

2.3 'Environment scanning' intelligence

Scanning the environment for specific phenomena may help companies to become prime movers and be at the leading edge in the business, or may sometimes identify high risks and emerging competition before it becomes a true threat. Constant monitoring of the surrounding world could be regarded as a 'company radar'. In modern shipping such as an ocean liner, the trained staff program the ship's route into the modern navigation devices. While the captain can be reasonably certain that the route and passage itself will be safely followed, there is no guarantee that another vessel is not in the way. Hence the radar screen is perhaps the most important equipment on the deck, is it not?

It is the management of a company who should pay keen attention to what is on their company radar screen. A well thought out set of topics that, whenever noted, should trigger some kind of alarm can give a tremendous advantage to one company. A small piece of news in a paper, telling that an electronic component factory burned down somewhere in the Far East, was not considered a major event in any of the media. However, the particular

Figure 2.3 How calling patterns change with age (iSociety March 2003)

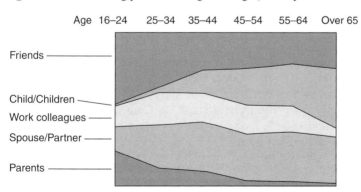

factory produced almost half of the world's supply of components needed in WAP-enabled mobile phones. Some managers in Nokia heard about the incident and immediately flew over to meet the other major manufacturer, wanting to make an exclusive purchase deal of their entire production. Nokia got a deal with the other supplier, while its main competitor, Ericsson, had to learn their lesson the hard way. Rumours tell that Ericsson could not introduce their WAP phone in time because of this incident, while Nokia was only little late with their launch and — most important — way ahead of Ericsson. We're not saying that Nokia had it's company radar working, but this merely illustrates the potential of how dramatic the impact of a weak signal can be in extreme cases, and that a quick response to a common threat may turn out to be a great opportunity.

2.3.1 Resource market intelligence

The above example of the burned down manufacturer might fall into the category of resource market intelligence. To some businesses, some resources are more important than others; it may be raw material, skilled labour, financing or good transportation infrastructure. For a brewery, a supply of clean and pure water is essential; for a rapidly expanding technology firm, recruiting a skilled workforce may become a real bottleneck and hindrance to growth. Some major projects, like building a nuclear power plant or a series of luxury Caribbean cruisers, may run into financing difficulties not because the banks won't lend the money, but because they are not willing to undertake the hedging (eliminating, for example, currency fluctuation risks by issuing currency futures). A shipyard in a small country may have to work quite hard to find a syndicate of several banks worldwide to secure the contract price and down payments over several years until the vessels are delivered.

In one example, a global-player telecoms operator planned to enter new markets by gaining significant ownership of a less developed telecommunications firm based in the Middle East. The scarce resource in this case was not the financial markets, the labour markets or know-how. The problem surfaced only when the technological platform of the acquired company was being modernized. It appeared that certain crucial components of the technology were listed by the US authorities as items under export licensing, i.e. not allowed to be exported to the country in question because of US security concerns.

Resource market intelligence usually looks into things important for the long-term plan, but sometimes things may change quite suddenly. A close down of a clothing factory will almost immediately release hundreds of low-salary workers in a particular geographic region. That workforce could be trained to do some other work for our company, or our subcontractor.

2.3.2 Reference market studies

Reference market studies are a very convenient way of forecasting the behaviour of own local markets. Although values and beliefs — the cultural background — affect our behaviour quite a lot, there are some success stories about how certain reference market studies have helped some companies to great success.

Yves Rocher — a cosmetics giant — applies certain methods in forecasting the demand of individual products in the Nordic countries. Finland, Sweden, Norway and Denmark are one another's reference markets. Once new products are introduced, some 'gift' item is bundled with the products. The key questions are how many orders do they get, and how many gifts should they order from the subcontractors. Some of the products with gift item X will be launched and introduced in the Swedish markets, while other products with gift B are introduced into Finish markets. Three to four weeks after the first advertisement campaign, they can forecast the demand for all new products in the launch markets with unbelievable accuracy. Furthermore, they can make remarkably accurate estimates of the demand for products in the other markets as well. This is, of course, a method that is constantly fine-tuned and of course, some odd outcomes occur every now and then.

In other businesses, we can learn from more mature markets and apply what we have learned in the less mature markets. We would like to point out that we should learn from mistakes in the mature markets too, not only from success stories. When it comes to mobile operator business, it is clearly an advantage to look what has happened in the Scandinavian markets, Italy, and now Japan and Korea. Which are the most successful services; how did the best companies introduce the services; what did the runners-up do differently, and why?

2.3.3 Partnership intelligence/networking

In telecoms many still wonder why 'partners' should be sought — can't we just settle for more inexpensive 'suppliers'. Networking — what's in it for me?

There is a global trend for companies to focus on the areas they are best at, becoming 'centres of excellence' in their own core businesses. Companies are focusing on their most precious and competitive assets, whether it is a location, human or other intangible asset, or any other factor or process that differentitates them from their competitors. We have, and we are now experiencing, a popular trend of outsourcing everything that does not increase value in the internal business processes of a company. Why is that?

Companies are constantly looking for suppliers whose business strategies are in line with theirs. A 'partner' who commits to growing together with our company in gaining a competitive edge in our markets is highly valuable. The 'requirements' are usually described as *overall quality, reliability, cost efficiency (=competitive prices), resilience and ability quickly to adapt to changes in the business environment, accurate and punctual deliveries, technical support or any after sales support, advanced technology*. This list is by no means exhaustive.

The essential problem to solve is: 'How to build the "tuned-up" value-creation-chain?' This is where, for example, the telecoms operator plays a crucial role. Possible approaches include building partnerships with service providers — they may be the ones also producing the technology. Since the mobile terminals, mobile phones, PDAs, lap-tops, etc., are specifically designed for communication, transmitting, receiving and storing relevant information, the selection of third-party content providers may become extremely important. We will discuss the finer points of partnerships in Chapter 5 later in this book.

As discussed earlier, nobody really wants to receive 'spam' or 'junk mail'; in mobile terms that would be anything with low relevance. The display units are not very practical and the price to pay for a high volume of irrelevant information forces the operators to focus on services that are either highly personalized (the relevance and value for the user exceeds clearly even a high price and inconvenience of sometimes bad display), or high-volume, low-price, services. There is also the risk of getting bad publicity or bad-will (as opposite to a good reputation) by partnering with badly performing service/content providers.

The end-user does not necessarily differentiate the initial service/content provider from the telecoms operator. Imagine a Spanish person regularly travels to France. The Spanish operator A has a roaming agreement with a small operator X in France. Assume the location this person visits is on the outer limits of X's wireless network. If the line is bad and connections are not reliable, this person is probably thinking that the A's network is poor, as his colleague using another operator and roaming through another stronger network does not experience similar problems. OK! No problem when there are very small number of people affected, but how does it work if the French operator has poor connections in the Riviera or other popular holiday resort — then the problem is likely to cause some adverse implications.

On the other hand, why settle just for building good relationship with selected 'suppliers'? Good relations with 'second tier' actors in our business is as important as the direct ones.† The word 'networking' is used to describe the horizontal (or second tier) co-operation while 'partnership' is meant to describe more the 'vertical' co-operation or the supply chain management, not to be limited only to that though.

Networking can be simplified to an equation $1+1=3$. A joint venture creates higher business value for the client as compared with any part of them alone or separately evaluated. It's a kind of the 'catch-22' backwards: everybody wins. To be able to find the right kind of 'non-competitive cooperating partners' (=joint venture partners) requires an extensive network of contacts and relatively good understanding of each networking-partner's business, as well as of their strengths and weaknesses. Trade fairs, associations and other similar communities are great places to meet the right people in the right place. It is a major effort and sometimes goes under PR, as that's what it essentially is. Of course this is one of the major responsibilities of upper management. Managers are usually experienced in communication and interpersonal skills, which is important, but we think the more levels of organizational hierarchies are involved the better.

2.4 Towards a higher intelligence

This chapter has examined market intelligence and the next chapter will look at customers in more detail through segmentation. The rest of the book will discuss how to engage in marketing actitivies. Those activities should never be entered into without a solid understanding of what the customers want, and what the current market situation is able to allow. Market intelligence should always guide marketing activities.

---

†Direct actors refer to competitors and suppliers and customers, 'second tier' actors refer to companies that are not exactly in our business, but whose success and actions are directly related to ours. For a telecom company a second tier actor could be handset manufacturer, or a financial institution providing either financial services or industry analysis reports.

It is, therefore, vital that market intelligence is used and trusted. The market intelligence units must retain impeccable standards of objectivity and deliver facts without taking sides in internal politics or pet projects. In this we are guided by the American President Woodrow Wilson's thought: 'One cool judgement is worth a thousand hasty counsels. This thing to do is to supply light and not heat.'

Ungexoshe Mpalambili. (You cannot chase two antelope at once)
— Zulu proverb

3

Segmentation
Understanding Your Customer

Segmentation is the key to satisfying customers and to optimal tariffing. Thus segmentation is the key both to market share success and to profitability. Segmentation is a tool that was not needed when the mobile telecoms industry was growing at high rates in the 1990s. Now, as competition gets tighter, growth is no longer achieved by new subscribers but rather by stealing customers from the competition: the time for segmentation has arrived. Beyond the idea of capturing customers from competitors money will increasingly be made by selling new or innovative services to existing

---

3G Marketing: Communities and Strategic Partnerships Tomi T. Ahonen, Timo Kasper and Sara Melkko
© 2004 John Wiley & Sons, Ltd ISBN: 0-470-85100-7

customers. Segmentation also is about getting to know your customer and thus detecting new potential, and through listening to customer comments to discover areas for innovation.

Many marketing books have been written about segmentation either with a general focus or concentrating on how segmentation can apply to a particular industry. Most segmentation theorists in other industries have been restricted to what limited data their industry can generate on customers. The telecoms industry has access to the most deep and complete information of all on the behaviour of customers in any industry. This wealth of information is beyond the comprehension of most segmentation experts with experience from other industries or academia. That is why segmentation for telecommunications needs to be addressed in its own chapter. This chapter will look at segmentation specifically for the mobile operator with a focus on new services and 3G.

3.1 What is segmentation?

Segmentation is the marketing process of grouping customers for the purpose of targeted marketing activies such as promotional messages, service development and pricing. Segmentation usually involves building a segmentation model with identified customer groups — segments — and may involve targeted marketing activities, which may be formally defined in segment marketing plans and segment marketing programmes. Various customer segments may be grouped into clusters for the purposes of marketing activities. Clusters may include customer segments that are similar — and often in complex models would be depicted next to each other — or dissimilar. Clusters simply mean that a number of customer segments have been grouped for any number of possible reasons. The customers within any given segment have similar characteristics and thus can be expected to react in similar ways to any marketing campaigns. Customers with varying characteristics may react in a similar way, thus clusters can include customers from multiple segments.

Segmentation is a most powerful tool if the model is built with good insight and data on customers both real and potential. For segmentation to succeed, each defined customer segment has to be 'discrete'. By discrete we mean that there exists some definition or definitions by which any one customer belongs to only one segment. There cannot be a situation where one customer could belong to either of two or more segments *at any one point in time*. It is not uncommon for customers to move from one segment

Segmentation

Figure 3.1

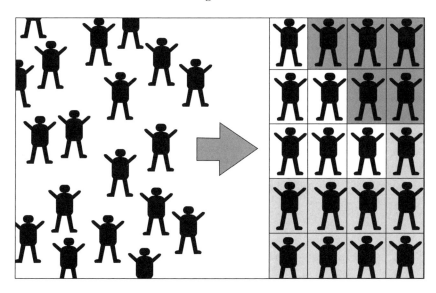

to another for any variety of reasons, such as an individual user's mobile phone behaviour changing when getting married or changing jobs, etc. Any segmentation model where customer definition is not discrete provides poor guidance to any decision making.

For segmentation to succeed the full segmentation model must define all existing and potential customers. Any prospective customer has to fit somewhere in the segmentation or else, again, the model provides poor guidance for marketing activities. Thus a good test of any potential segmentation model is to take a few real customers and see if everyone of them fits somewhere within the model, each customer fitting into one, and only one, segment.

3.1.1 Test of current telecoms segmentation

Most mobile operators will say they use segmentation already. As precise customer understanding was not a major goal of mobile telecoms — during the last decade the main goal was keeping up with the demand of rapid growth — segmentation used by most mobile operators tended to be limited, simplistic and, unfortunately for current needs, very inefficient. The most common segmentation models in telecoms are (i) segmentation by size, (ii) segmentation by technology, and (iii) segmentation by billing. It is also not uncommon to have a model mixing two of the above segmentation criteria. Let us take a quick look at these three common models.

3.1.1.1 Segmentation by size

For example, large corporate customers, SME (small and medium enterprise) customers, SOHO (small office home office) customers, and residential customers. Note that this example model will therefore have four segments. Segmentation by size correlates very poorly with mobile telecoms service usage. A five-person taxi company will use cellular services in a similar way to a 200-person taxi company. However, both taxi services consume telecoms services in a totally different way to a printing house, whether that printing house employs five people or 200.

3.1.1.2 Segmentation by technology

For example GSM, WAP and GPRS customers — a model of three segments. The technical ability of a handset is a limiting factor in our industry but provides little if any guidance as to how the customer may behave. A 50-year-old lady with a two-year-old GSM phone is likely to behave in a very similar way to a 50-year-old lady with the newest phone. However, the newest phone is definite to have WAP and can very well have GPRS (or on CDMA systems 1×RTT). There is likely to be much better correlation with the age than with the technology of the phone in the case of the 50-year-old woman. To show how poor technology is as a predictor, contrast the phone, SMS and ringing tone purhase behaviour of two 17-year-old boys with the two ladies mentioned above. A 17-year-old boy with a GSM phone would behave much more like another 17-year-old with a GPRS phone rather than mimicking the behaviour of the 50-year-old lady with the GSM phone.

3.1.1.3 Segmentation by billing

For example prepaid (voucher) customers and post-paid (contract) customers. Note here we have a segmentation model of two segments. Again, a 24-year-old single mother is likely to behave similarly to another 24-year-old single mother, regardless of whether one is on a prepaid and the other on a post-paid account. Neither would be very similar in phone use to the jet-setting executive who might be on prepaid or post-paid.

If your model is based on any of the above grounds, recognize that you are using an instrument that is useless for modern mobile telecoms use. It is like trying to use a hammer to strike a screw into a wall. Yes, something may come out of it, but nothing lasting or worthwhile. The worst aspect is that many in your organization may be confused into thinking that they are using a practical tool — after all it is simplistic segmentation — and end up blaming marketing, the segmentation, and rejecting modern scientific marketing methods.

3.2 Segmentation criteria

Remember that the purpose of segmentation is to guide the targeting of marketing activities. A customer base can be segmented in numerous ways and no one segmentation is perfect for all situations or for all operators. As we discussed before, the size of the customer, the technology of their handset(s), and whether the customer is on prepaid or post-paid account, are all pretty much meaningless as segmentation criteria. This section will next examine other criteria, most of which are much better than the discarded 'Top Three' most used segmentation models.

3.2.1 Segmentation from the academics

In most other industries there is very little actual end-user behaviour factual data. For example the Ford Motor Company does not know which of its customers opens the passenger-side door every day several times — and thus could be expected to have two people travelling in that given car — and which drivers do so very rarely and thus likely to have only one person in the car. However, in telecoms the mobile operator knows every time we use our phone, and also knows the location of our phone, every call made, every service used, every button pressed and who we communicated with. Companies in all industries spend a lot of time and effort conducting surveys, focus groups and other analysis in attempting to discover how their services and products are used. When segmentation theorists have developed more useful segmentation tools, they have tried to identify various patterns and these tend to arise from sociological behaviour models, etc. These may be useful in telecoms, but do not capitalize on the powerful information that is inherent in this industry and few others. As numerous telecoms segmentation conferences feature these types of discussions, let us cover them briefly.

3.2.2 Segmentation by geographical pattern

This approach assumes that an environmental factor is determining decision-making patterns. This is, of course, true to some extent, for example a company selling boats is more likely to attract business from people living along a coastline like Miami or Rome, than from those living further inland, like in Las Vegas or Vienna. On a smaller scale, the environmental or geographical factors as far less accurate; for example, how many identical cars, pet animals or hobbies do you think people have who live within the same ZIP-code area with you? Segment by geography is not an ideal tool.

3.2.3 Segment by demographics

Although the greek word *demos* means 'people' the following applies as well to business customers. Demographic data includes age, gender, level of income, address, etc. This is data that is usually readily available from government statistics. The demographic approach assumes that the key determinant for decision making follows a certain pattern based on a demographic criteria. For example all 25- to 30-year old single males are the most likely to buy a six-pack of beer before every Tuesday night basketball game. In most industries this is a common basis for segmentation, for example aiming certain electronic gadgets at young men or a certain magazine is aimed at married women living in cities, etc. As there is remarkable variety in all societies among people in any given demographic group, especially when it comes to consuming digital content and managing communication needs, demographics prove to be of very limited value. In segmentation in telecommunciations, demographics are only marginally useful, primarily only for new operators if no real customer data is available.

3.2.4 Industry type

Business customers can be classified by industry. A computer networking company with many field engineers installing, updating and repairing computer networking equipment is likely to have similar mobile telecoms needs to another computer networking company, and both of those quite different

Figure 3.2 Vertical Markets will be providing much of the early mobile service revenues in 2002

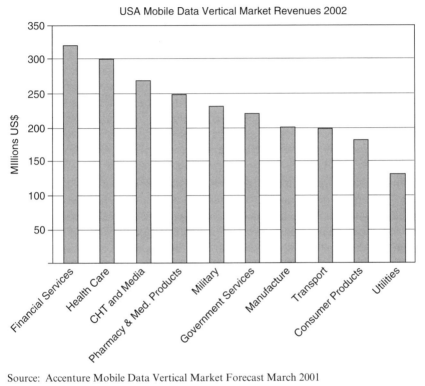

Source: Accenture Mobile Data Vertical Market Forecast March 2001

from the telecoms needs of a clothing manufacturer, where most employees are working beside sewing machines. Industry type can yield good insight into telecoms behaviour, but a standard industry classification is likely to produce artificial differences where one industry is certainly quite different from the overall economy point of view, but would be similar in mobile telecoms usage. In the above example, a florist delivery service may have very similar mobile telecoms behaviour to the data networking company, yet most industry classifications probably place flower deliveries and computer networks very far from each other. A lot of judgement is needed to find useful groupings of industries.

3.2.5 Segmentation by using various distribution channels

A company typically used to fall into one category, either as being in the *distribution channel* business or selling its products or services *through a distribution channel(s)*. A traditional airline company is a suitable example of a company selling its services through a distribution channel. It usually has large number of travel agencies in several locations to sell flight tickets for its flights. Any wholesale business is a perfect example of being in the distribution channel business itself — it does not produce anything tangible and the added value it creates is very hard to measure, except through standard financial performance indicators at the end of an accounting period.

3.2.6 Personal data

Some of the information that could be used are various personal data such as age, gender, marital status, etc. These may provide some guidance to customer behaviour. For those operators who do not have a customer database — i.e. new entrants with no existing customers — personal data may be a good starting point. But as we know, two 26 year olds may behave radically in different ways, and personal data is not a strong predictor of the type of telecoms behaviour that a given customer might produce. Still it is fair to say that young people are usually much more likely to try new things, women and men have differences in communication behaviour, and marital status etc can have predictive uses. This is not bad information for anyone in segmentation, but we can find much more powerful data to use from within the telecoms world.

3.2.7 Segmentation by psychological patterns

Understanding some unconscious factors that drive the decision making process can help and support definition of the most appropriate ways to reach different segments. For example, if a company wants to appeal to patriotic feelings among its customers in the USA, it would be obvious to use the colours blue, red and white — found in United States flag. However the same colour mix will most probably not have similar response in the citizens of Brazil.

3.3 ERP, CRM and segmentation

In the 1980s, the golden era of enterprise resource planning systems, the PC revolutionized the data processing capabilities of small and medium sized enterprises. 'ERP' became popular jargon later in 1980s. Soon the IT business moved its focus towards another acronym CRM — customer relationship management. In the early 1990s, the vast majority of large enterprises were planning to invest more money on CRM systems than on any other business process support systems. Several studies have shown that retaining a customer relationship is only about 10–20% of the costs of winning a new customer. This finding has fuelled many customer loyalty programs — some that are successful and others less so — all aiming to improve churn and competitive edge. Soon the competitive edge deteriorated as loyalty incentives became very complex and the benefits were really hard to quantify — at least for ordinary consumers. The CRM investments are delivering diminishing returns.

CRM systems were designed to keep track of sales-force activities, customer feedback and other factors that could be used to serve customers better or achieve better satisfaction. CRM systems can be used to classify customers into homogeneous groups or segments, and to identify correlations in behaviour. A good model should then provide predictive outputs, i.e. a customer with a given profile is expected to become an early adopter of a given type of service.

When segmenting, you have to be careful not to think too narrowly. It might happen that business customers of small office home office (SOHO) type behave similarly in some cases to residential customers, etc. In some cases heavy users from the residential customer base might behave much more like a business customer than another average-use residential customer.

Segmentation as such is not a new concept. The marketing strategies of, for example, some masculine products — like sports cars and cigarettes — have been designed so that a certain part of the great public — a target group of consumers (a segment) — will find it easy to associate to or relate themselves to the values and image that the advertisement passes on and communicates. The message and image have often been emphasized by having well known celebrities and other famous persons — as a sort of role model and idol — appear in the ads.

3.3.1 From hard to soft facts

Segmentation is not only about marketing existing services to existing customers, but also about really understanding the customer and his needs and, in consequence, find new product development ideas. Operators clearly have a competitive edge in collecting customer information from the technical billing and charging data. They also have indirect access to some softer facts and knowledge through their high customer commitment and proximity to users.

Segmentation information and customer contact can be enhanced, for example, via the phone bill. Operators can be sure that the billing letter is always opened. The bitter pill of paying the sum on the invoice can be sweetened by showing how many customer loyalty programme points the customer has accumulated so far, etc.

Customer loyalty programmes could reward users to acquire new customers, answer questionaries, or fill in/update their customer profile. As stated before, the collection and updating of soft facts are the true key to success. Really getting to know different segments is more about psychology and sociology than the 'hard' billing data. This is why companies should keep trend studies in mind in order to develop new product ideas.

A professional marketing organization uses resources and energy to monitor the behaviour of their customers. Behavioural information to be tracked can include:

- percentage of business/private use;
- communication with family vs friends;
- types of different communication to different times of day;
- scheduling (facts) vs entertainment (fun);
- active (calling, sending SMS/MMS) vs passive (receiving calls, SMS, MMS) vs avoiding (voice mail) communication.

3.3.2 Users broken down — segmenting situations

Mobile behaviour, as with many other sorts of behaviour, tends to be a hybrid, i.e. behaviour changes according to different sociological roles during the day. The use of modern communication media differs greatly between work and leisure. One might think that some VIPs are allowed to reach you at any given time of the day. In reality, there are situations where you prefer not to be disturbed at all — even by your spouse. The form of communication and the motivation is thus dependent on your current role, time, and even your feelings.

The consequence is that segmentation is not limited to people but can cover situational conditions, too. Marketing efforts could be adapted to the time of day and the role you are in. We claim that traditional segmentation models are far too limited for modern communication needs.

3.4 From theory to practice: building a segmentation model

We have examined why segmentation is needed. We have also covered briefly several ideas about segmentation from outside the telecoms world where companies tend to be rather 'blind' about their end-user behaviour. Now let us turn to the practical issues of using segmentation in marketing activities in mobile telecoms where operators have 'perfect' information on end-users. What is needed is a modern, *powerful* segmentation model.

3.4.1 Characteristics of a useful segmentation model

A useful segmentation model has customer segments which are heterogenous without, and homogenous within. In other words all customers within a segment are similar, and customers between segments have significant differences. Let us start with the phrase significant differences. We could devise models that classify customers by clearly differentiating facts that do not assist in targeting telecoms services. For an extreme example, imagine segmentation by the colour of the eyes of the customers. Blue-eyed customers would be clearly distinguishable from brown-eyed customers, but most probably eye colour does not correlate well with mobile service usage. (Note that this is not an impossible segmentation criteria: eye colour could be the basis of very valid segmentation in some other industry, for example the contact lens industry.)

The aim of segmentation is to give tools for targeted marketing. To accomplish that, a good segmentation model is built around customer characteristics which have a significant impact on the usage patterns for the intended services. A good segmentation model should explain why any given segment would behave differently if faced with some promotion, price, service, etc. Also, the model should help predict behaviour to help target our marketing efforts.

3.4.2 Segmentation by user behaviour

Telecoms operators can go much further in using the potential power of segmentation than can any other industries because of the depth of information gathered automatically about every user on the network. In the automobile industry, car manufacturers do not know how many times a given driver opens the boot, or how many times the car is driven over a bump every day. Some information in the car industry can be gathered from user surveys, or from examining data collected at maintenance intervals, but telecoms is the only industry where the operator knows every single time that the user presses any key on the keypad of the mobile phone. Fantastic levels of user information are automatically gathered and can be developed for segmentation based on user behaviour. As no other industry has this wealth of information, and as telecoms itself did not need sophisticated segmentation in its rapid growth phase in the 1990s, this is very much a hidden asset.

The actual segmentation based on user behaviour will, by definition, differ for every operator as the actual behaviour of users differ across countries and within markets. Some very basic early ideas for tracking user behaviour, and building powerful segments, include tracking which customers (i) inititate calls and who tend to only receive calls. Some customers show a clear preference to (ii) sending SMS text messages while others prefer voice calls. Mobile telecoms data is incredibly deep. Not only do we know how many minutes every customer places phone calls, we also know what kind of call. For example some place only a (iii) few calls per week but these tend to be long calls, while other customers make many calls per day but each is of short duration.

Figure 3.3 Some people are more connected

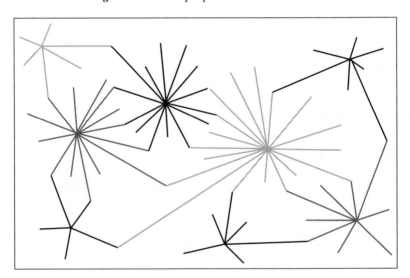

Segmentation 43

Some customers place calls (iv) during office hours but almost none in the evenings or weekends, which suggests that this customer may have another phone on another network for evening and weekend use. Some customers perform (v) return calls, such as the child calling the mother's phone and the mother returning the call, so that the phone call is not billed to the child's account but rather to the parent's account. Some callers have a very tight set of only a (iv) few people who they communicate with, while others place and receive calls to and from seemingly random sets of numbers. Such behaviour-based patterns can be identified and then customer segments can be built.

We want to stress that true behaviour-based segmentation is *infinitely* more powerful than any of the traditional methods mentioned earlier in this chapter. Imagine two 32-year-old women living in the suburb of a major city, both housewives and mothers of two young children of toddler age. Let us also assume both have a WAP phone on a post-paid account. By any other segmentation method they are likely to be grouped into the same segment, whether by age, location, technology, billing system, customer size (residential as opposed to business), etc.

These two women may behave in a similar way but one might suddenly start to behave in a very different way. If for some reason one woman becomes estranged from her husband — and the two take a trial separation, which of course does not yet alter her marital status — she might start to exhibit mobile phone behaviour very similar to 17-year-old girls with their mobile phones. She could start to send lots of love notes via SMS text messaging, perhaps even chatting on the mobile phone, and possibly using dating services. This is a typical behaviour pattern of people who have recently become single. They originally did not learn to use the mobile phone in the dating situation, but now will rapidly pick up habits very similar to those associated with older teenagers. The mobile operator needs to observe the *behaviour* of the customer, not blindly place the customer into a demographic category and assume that woman will continue to behave exactly like her happily married housewife neighbour.

3.4.3 How many segments?

At the one extreme a mobile telecoms segmentation model might have two segments — business and residential customers. There are likely differences in behaviour between business customers and residential customers, so this model could provide insight, even though it only has two segments. At the other extreme, many modern marketing theorists have argued on behalf of segmentation developing quickly with advanced data mining and data processing ability, combined with vastly improved data on customers, so that marketing could be targeted even to the detail of a 'segment of one'. Some have gone as far as to suggest that a segment of one is a desirable and even practical tool to bring truly focused marketing offerings. In some industries with very few global customers, the whole potential customer base can be separately defined and a model with each customer being its own segment can be used.

For mobile telecoms, with subscriber numbers in the millions, a model of only two segments provides very little practical illumination on what marketing efforts to direct at any given customer. The more segments in a model, the more complex it becomes to build, test, train users on, update and refine, and use. While a million segments — approaching a segment of one — seems like remarkable overkill in the current state of telecoms competition, a good benchmark guideline has emerged. Nokia has been quoted in the public press, such as the *Economist* in January 2002, as having an end-user segmentation for its mobile phones, which has 35 defined segments (a two-dimensional grid of 7×5 according to the article). Nokia is reportedly addressing about half of those, probably for market size and competitive reasons, leaving some of the smaller or less profitable segments to its competitors.

Without suggesting that Nokia's model is in any way the 'best' for mobile telecoms, it does provide interesting insight to modern mobile operators. As Nokia does not sell its handsets directly to the end-users, and uses the mobile operators as its distribution channel, the Nokia segmentation model is a good benchmark for considering the customer of the mobile operator. If Nokia already divides all end-user customers into 35 segments, then any model used by a mobile operator which has less segments, has by definition less precision and is thus actually *weaker than that of its supplier*. A mobile operator should be very concerned if its own marketing is directed by less precision than that of its equipment supplier, whose products go to the operator's customers. If the mobile operator wants to capitalize on its intimate knowledge of its customers, then that know-how has to be analysed well enough to produce at least a model of similar depth as the Nokia model, preferrably better.

We would argue that a practical segmentation model for a mobile operator today has at least one degree of magnitude — at least one dimension if you will — more than that which Nokia reportedly has. In other words an

Figure 3.4 Two- and three-dimensional models

Two-Dimensional Model 3×4

Three-Dimensional Model 3×4×4

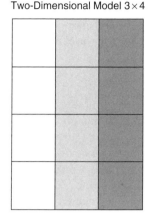
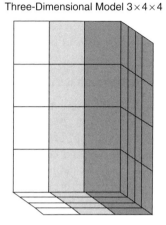

'adequate' segmentation has at least 100–150 defined segments, so that any customer fits only one segment, and no customer is outside the model. The mobile operator does not have to provide services or marketing activities to every defined segment, but each segment must be defined and at least superficially analysed. For a mobile operator with about 10–15 million subscribers, every segment would then be on average of the size of 100 000 customers, allowing considerable precision in targetting individual segments and clusters of segments.

One such three-dimensional model could be for example with segmentation dimensions (i) who pays the phone bill; (ii) evolution on the usage of new mobile services; and (iii) type of networking activity. An employee who gets free mobile phone services behaves differently to a person who pays the bill, who again behaves differently from a child or spouse whose family member pays the phone bill. The evolution of user means the user doesn't use SMS, or only receives SMS, or sends SMS, or uses m-Commerce, or sends MMS. Types of networking activity could be divided, for example, into avoiding contacts (lots of missed calls and voice mail), many more received than initiated contacts, low initiated contacts in general, many but random contacts, or many regular contacts.

The above, still simple, model could have 4×5×5 segments, i.e. 100 in total. One segment would be the person whose employer pays the phone bill, who sends MMS and contacts many regular people. This is an ideal segment to target new services and is very likely someone who influences other people. Another segment would be the user who pays his/her own phone bill, only receives SMS and has low initiated contacts. This person is likely to be ready for a campaign of creating the addiction to SMS and could be offered a few months of free SMS to get the user hooked, and so on.

3.4.4 Comparison with the car industry

When you consider going from what may be four segments today into something between 100–200 segments in a new model, the first reaction tends to be that such an exercise is time consuming, expensive, hopeless, overkill, and that it would be much too complex to be of any use or utility and the whole model would be abandoned as nobody would use it. These are very typical responses, especially of the majority of the telecommunications senior staff who are very much engineering oriented and technically trained. They do not see the inherent benefits of a major segmentation project, especially if the current segmentation model is seen as adequate and/or not particularly useful.

To understand by simple example how powerful a powerful model can be, let us examine the automobile industry. The first Model T Fords were made to one model type with one engine and one colour — black. While Ford concentrated on maximizing the manufacturing utility of efficiency the production line, other manufactures like General Motors started to develop segmentation to guide the development of their lines of cars. Eventually even Ford had to adopt segmentation. A family car emerged, as did a youth oriented inexpensive sporty car, a luxury car, and a utility tool (light truck). As the customer understanding grew, so too came more precise model development and refinement. Now there are segmentation models that have defined dozens of classes within those broad areas — and of course introduced many new segments as well.

The product development (car design), pricing, promotion, distribution, sales pitch, etc., are all tailored by segment. For example, some car is clearly targeted at women, and a variation of that car to mothers with children. The advertising is aimed at TV and other media that the identified target audience views. A sales representative is trained to mention the expected best arguments and attributes for any given segment, so when the sales representative finds out that the woman who walked into the dealership is a mother with children, the sales pitch for the car is different from the sales pitch to a woman of the same age who is a single professional.

While a segmentation model may have a lot of detail and depth, it does not have to be overwhelmingly complex. A technique to manage the difficulty and complexity of the model is to use dimensions. In the above automobile example the basic level segmentation could be by gender – men and women seem to have different tastes in cars and respond to different arguments when selecting a car. The next level (dimension) could be age, for example, and another could be marital status. Thus a new employee could be shown the model at its high level, dividing customers only by gender and age, explaining why a 55-year-old man is interested in very different arguments for the same car as would a 22-year-old woman. Then when the employee wants further understanding, further dimensions of the model can be explained such as marital status and whether the customer has children, bringing increasing depth to the understanding. A large store would have sales staff that specializes in some types of models — and customer segments. In a similar way the segmentation model in telecoms can have several dimensions and, at its top level, the first two dimensions could have from 20 to 40 segments. Then added depth from further dimensions could split the segments further yielding hundreds, even thousands, of segments.

Again we must caution against using outside demographic data, but rather *real* user behaviour data. Only in telecoms do we have the luxury of knowing everything our customers do with their phones within our network. The best segmentation is based on that unique and powerful insight.

3.4.5 Beyond a segment of one

Traditional segmentation has the concept of a human individual being the smallest unit that can be assigned to a certain cell in the segmentation system, but why don't we go further than that? In telecoms segmentation we can go well beyond the so-called segment of one. Sociologists and human behaviour scientists have proved that people behave in different ways at different times of the day. We exhibit behaviour based on what role we play. With telecoms segmentation we could break down the user behaviour into the different role characters. For example, a professional woman worker may also be a mother, and depending on whether her boss or her child calls, she will behave very differently, and do so at different times of day, etc. With this insight we can find that that otherwise very different users that belong to totally different segments can have the same service needs, depending on their current situation (time of day, leisure or business, etc.).

Why not cross-segment situations and roles with users? For instance, the typical single business woman on holiday with her girlfiends; the adolescent in school; the family father at work. Then break down the people in the situations to their basic needs — what proportion of communication is entertainment, what means of communication is preferred because most fun or most efficient or cheapest, etc., reachability for whom, etc. Understanding the situation and the needs means that marketing actions can be targeted more accurately.

3.4.6 From business to individual

One has to keep in mind that behind business customers there is always the individual as end-user. The acceptance of business services reflects on private behaviour and vice versa. Once again, the buyer or user is hybrid — a business customer is also private at the same time. A powerful segmentation model will incorporate this insight as well. A segmentation model will need to be developed with the best insight of customer understanding and behaviour, ideally with marketing managers working with sociologists, behavioural psychologists, etc. Some may be employed in-house while others may be hired from specialist firms that study behaviour especially relating to mobile telecoms.

3.4.7 Self-organizing maps

At some point the utility of employing psychologists and sociologists with the marketing managers and segmentation managers to refine and redefine ever more insightful segments will become too costly to justify it. At that stage telecoms has to turn to the science of applied mathematics. Modern tools exist to define microsegments on patterns that are not intuitively obvious, but which can be identified by mathematical means using neural network technology. The marketing segmentation applications are called self-organizing maps (SOMs, Figure 3.5). The limitations of traditional intuitive segmentation models — like the examples discussed here — stem from the limitations of the segmentation managers in combining marketing, sociological and psychographic knowledge. The depth and utility of the segmentation model will be dependent on how well the segmentation managers understand the business and the customers, and how well they can define and describe these. No matter how brilliant, eventually any segmentation manager will exhaust the ability to develop ever deeper insight. Also, it takes a lot of time and effort to build the know-how to evolve the segmentation.

Segmentation

Figure 3.5 Self-organizing map (SOM), grouping customers with similar characteristics (Xtract Ltd)

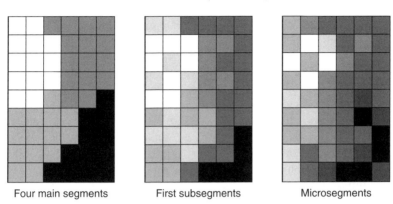

Four main segments   First subsegments   Microsegments

Self-organizing maps ignore the labels and sociological 'reasons' for why people behave in a given way, rather they simply identify and isolate groupings of customers who behave in a similar way and group them separate from those with maximally different behaviour. The SOMs just analyse massive amounts of data and group them by relevance, producing powerful segments that may at first glance appear random or coincidental. This type of segmentation identifies the correlations between various criteria, often as a combination of several seemingly unrelated items. It is up to marketers to analyse the results, discover the reasons and give labels to emerging microsegments as needed. Such mathematical models are still in their early stages but already the first commercial applications have emerged both in telecoms and in the banking world. Looking at a future of telecoms with subscriber numbers in the tens of millions, and the need rapidly to identify new microsegments for rapidly emerging and evolving services, mobile operators and their partners should invest in new mathematical segmentation systems to gain competitive advantage. No matter how clever any data-mining solutions can become, self-organizing maps are inherently superior due to their automated nature of discovering the relevant differences.

3.4.8 From alphas to omegas

Current leading-edge neural network technology has started to identify certain groups of individuals within the telecoms customer base that are worthy of exceptional attention by the mobile operators. Using SOMs, operators can identify for example alpha users (Figure 3.6). Alphas are those amongst us who are very well connected, who initiate a lot of voice calls and SMS and MMS messages, and who seem to be keeping everybody informed about meetings, parties, clubs, the changes of schedules, etc. Alpha users are network hubs or concentration points connecting human networks. An alpha user may have a very broad range of human contacts that the alpha person happily sustains. Alphas therefore are exceptionally attractive targets in cultivating telecoms user communities. When launching a new service, those alphas that are likely to find the new service appealing should be the initial target of the launch campaign. These can be very cost-effectively targeted, and through a very limited number of alpha contacts, the whole population of interested people can be reached.

It is important to note that just counting contacts in personal networks will not yield optimal Alpha user groupings. Some communities have several parallel hubs, and targetting those will not yield any improvement in contants, as these highly connected hubs connect to the same network. It would be a waste of marketing effort. Other Alpha users are hidden as hubs of autonomous but smallish networks. Their connectedness may be at or even below average, but as they are the hubs of otherwise autonomous networks, they still would deliver significant user groups as hubs.

Many other such findings of network and community groups can be discovered using SOMs. For example omegas — the people who do not

Figure 3.6 Role of alpha users in service adoption (Xtract Ltd)

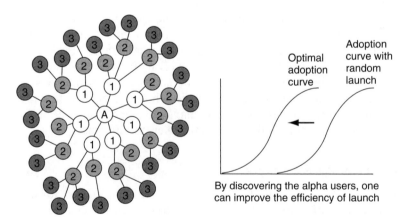

By discovering the alpha users, one can improve the efficiency of launch

Segmentation   51

initiate calls, but who receive a disproportionately high amount of voice minutes — are key to profits for most mobile operators, as the money made on terminating calls to these subscribers is very profitable traffic. However, traditional churn management systems looking at high billing data will invariably totally ignore the omegas and inadvertently dismiss them as irrelevant. Only SOMs can efficiently discover such customer groups.

Still another set comprises those customers who have multiple subscriptions, on your network and with one or more of your competitors. Again SOMs and certain user behaviour patterns and profiles developed and then analysed from the monthly traffic data on the whole network, can identify these and even give a moderately good guess as to what the other phone number is and thus on whose network your customer is also a subscriber. These kinds of finding are being developed by advanced mathematics applications for telecoms segmentation, especially using self-organizing maps. In 5 years all operators will necessarily have SOMs in their arsenal and providing the 'long range radar' to support the more conventional data-mining tools.

3.5 Developing the segmentation model

Let us stay away from SOMs and return to our 'manual model' of a few dimensions. After relevant segmentation criteria are identified, a model can be built. A typical two-dimensional model would have the two most useful and predictive segmentation criteria. The two criteria would be placed into a matrix, yielding a model with cells. Thus the first decision in building a model is to prioritize the criteria. Additional dimensions can be added to the model to provide more information. These should be added again in order of predictive utility.

After a model has segments (cells) defined, each cell will need some data and analysis. Each cell should be given a provocative but descriptive name. The temptation is to call model cells A1, B2, C4, etc., but the utility of addressing a new sales campaign targeted at model cells C3 and D4 provides very little assistance to all who have to use it, and opens up much risk of confusion. Much more useful are descriptive names, which could be 'young adopters', 'single parents' and 'retired adventurers'. Now when we say that the campaign is targeted at 'retired adventurers' and 'working sportsters', we have a good understanding of who is our intended target audience.

Each cell in the model needs to be defined by the selection and exclusion criteria, to enable anybody using the model to determine easily which customer belongs where. The model should be so clearly defined and intuitive in its use and naming that no confusion exists even among those who are not trained to use the model. The definition for the young adopter segment could be for example 'under 25-year olds who have at least two of the following – PC, PDA, digital camera, minidisk player, i-pod, or GPRS/3G phone'.

Next a rough mapping of the overall market needs to be made against the segmentation model. So if the operator has 6 million subscribers, a rough analysis should be made of how many of the 6 million are in any given segment. The other prospective customers in the market also need to be mapped into the model. This split of the total prospect population is the first step to prioritizing model cells. It is quite typical that some cells are very large — in this example two cells might end up with over 1 million existing customers in each — and others are quite small, numbering thousands or even only hundreds of customers. It is not at all unusual that some cells of the segmentation model are even totally void of customers. That is not a failure of a model, but only indicates that some segments are only hypothetical. In such cases some consideration should be given to considering if the cells will have the possibility for customers emerging from that group in the future (Figure 3.7).

Figure 3.7 Migration paths for customers in a segmentation model. A powerful segmentation model may include predictive abilities for how customers evolve and migrate from one segment to another over time

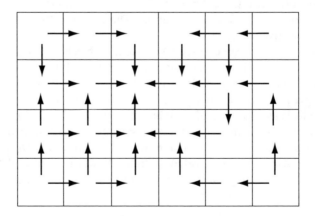

Segmentation

For example, if one of the dimensions of the model is age, and another is adoption rate, then the model would yield a cell where there are young users who are early adopters, but the model would also yield a theoretical cell of retirement-age users who are early adopters. It is likely that there are not very many retirement-age people who are still eager to learn ever newer technologies, so this cell could be very small in the population, and could even be zero. Management judgement is needed as to whether or not it is possible in the near future that this cell might grow to have some customers, for example after marketing campaigns targeted at retirement-age customers. Segmentation Model cell populations change over time, and individual customers migrate from one type of customer to another. It is important to remember that any one customer can be in only one cell at any given point in time — but the same customer may have been in a different cell or series of cells in the past and again in other cells at other times in the future.

After the prospective customer population is split into the segments, analysis of significant cells will need to be made. Not all cells need to be analysed — definitely not those that have zero population. Analysis such as the cell revenue size (number of customers in the segment times average customer revenue within the segment), revenue growth, competitor presence (which of your competitors serve customers in that cell) and your product offerings into that cell are to be collected.

As customer insight is developed, more sophisticated understanding can emerge, such as typical migration patterns. For example, it might be typical that if the model has age and employment as criteria, that a young person starts in the cell young and in school, then moves either into one of young and working, young and unemployed, or young and at university. As a young and working customer is likely to view mobile services in a more positive light than a young and in school customer, that migration — movement from one cell to another — can provide a significant marketing opportunity. As conventional wisdom suggests, any young person who leaves school to start his or her first job, would be a good candidate to sell any new services or products, as the disposable income takes a dramatic jump.

Ever further insight can be developed depending on which criteria are used and what data is available on the customers and their behaviour. Results from marketing campaigns, the profitability of services, the attractiveness of service bundles, the sales arguments that work, etc., etc., etc., can be used to bring further power into the segmentation model. It is important to note that at this level of depth, not all model cells need to be analysed. In fact the operator should select target segments and develop a deep understanding only of those where a significant marketing effort is intended. A key determinant is the competitor offerings in any given segment and the relative market opportunities for that segment.

Eventually, segmentation models should also track the communication networks of each user. Again these differ radically. Some mobile phone users call a lot, others only a little. Some contact many people, some only a few. Some build constantly new friendships and their networks evolve a lot. Others communicate with essentially the same group of people over years, decades even. Some initiate, some receive. These all factors should be mapped against behaviour and new types of services, to isolate early adopters and to help target marketing campaign.

3.6 To sum up segmentation

This chapter has looked at how customer insight is captured into a segmentation model, what is needed for powerful segmentation, and how the model can be built and refined. We again remind the reader that marketing covers everything from product design to its pricing, promotion, sales and delivery. When looking at the advanced mobile services arena of 3G and the vast numbers of customers, *segmentation is the only way* to deliver targeted marketing activities.

This is a book about modern marketing. Therefore, everything in this book must be based on segmentation. That also means that poorly designed segmentation limits the chances for the company to succeed, while powerful segmentation will deliver competitive benefits throughout the organization. We all of course admit that engineering is important, that looking after profitability is important, that keeping a constant vigil on what the competitors are doing is also important, but most important for the long term success of a company under conditions of competition is what the customer thinks. As Hugo Paulson said: 'Good firms worry about competition. Great firms worry about clients'.

No one can possibly achieve any real and lasting success or 'get rich' in business by being a conformist.
— J. Paul Getty

4

Service Development and Management
Building Desire

3G is not about technology, it is about services. The service creation environment for 3G will be richer and offer a greater range than any service creation environment ever before. The initial services that will emerge in 2.5G and 3G will tend to be mobile phone variations of services that exist on other media, such as the fixed Internet, TV, CD-ROMs, etc. As service

---

3G Marketing: Communities and Strategic Partnerships   Tomi T. Ahonen, Timo Kasper and Sara Melkko
© 2004 John Wiley & Sons, Ltd   ISBN: 0-470-85100-7

creation engineers discover the wealth of opportunity with 2.5G and 3G, the world will increasingly see more exciting and creative services.

This chapter examines how services are created and how added value can be built into services when they are brought to the mobile environment. This chapter touches upon service tariffing, which will be covered in a later chapter (Chapter 14). This chapter spends more time on exploring ways to ignite the creative impulses for services, and also discusses the way services are managed and, eventually, terminated.

4.1 Product development — the Five Ms

First we examine how services can be created for the mobile phone through the theory of the Five Ms. The Five Ms of mobile service creation are Movement, Moment, Me, Money and Machines. Movement, escaping the fixed place is not the same as location information. Movement includes location information but needs to understand how a moving user and the surrounding environment relate. In some cases we want to know something that is tied to where we are — such as finding the nearest cash machine or Pizza Hut. At other times we want to access a service that is anchored to where we live, such as accessing the stock market information of our home country, not that of the country we are visiting. Still other Movement services must travel with us, such as a map in our car being updated as we drive along. Movement is the distinguishing aspect of cellular telecoms that other digital service delivery systems cannot easily replicate. As such needs are discovered where the Movement attribute is strong, resulting services will be unique to mobile networks.

The 5 Ms
Movement — escaping place (local, global, home base, mobile, position)
Moment — expanding time (plan, postpone, stretch, fill, catch up, multitask, real time)
Me — extending myself and my community (personal, relevant, customised, community, permission, language, multi-session)
Money — expending financial resources (m-commerce, micro-payments, m-banking, m-wallet, m-advertising, sponsorship)
Machines — empowering devices (telematics, machines, appliances, robotics, automation, connecting with...)

Moment, expanding the concept of time, is the second attribute of the Five Ms. Moment is more than urgent services. Moment includes managing time, postponing time, catching up on lost time, multitasking, and hence 'creating' time as well as doing things with sudden extra moments of time or 'killing' time. The mobile phone has become only the second device that we carry with us at all times, but much more advanced than the wristwatch, the mobile phone allows connectivity and communication. Because we carry it with us at all times, the mobile phone is the fastest receptacle to use for any urgent information. This helps to explain why 80% of business executives in the UK already use SMS. The mobile phone can easily receive silent updates even in crowded meetings, in the theatre, bus, restaurant, etc. When speed is the key, the mobile phone holds an absolute competitive advantage over any other communication means. Lessons from Japan and the Philippines tell us that entertainment services on cellular phones are used to kill time. Adults would not buy a mobile phone to read soap opera or receive World Cup football updates, nor to read horoscopes and cartoons, but when they are available in your pocket and part of the basic mobile services package, we consume such entertainment just as we read the cartoons, horoscopes and sports updates in our newspapers. Moment is the most activating of the Five Ms and is the key to impulse purchases and acting on emotion.

4.1.1 Power of personalization

Me, extending myself into my communities, is the third attribute of the Five Ms. Me includes personalization, customization and all the ways to interact with our various communities. Me explains why the interchangeable covers are so popular with mobile phones, and why so many feel very strongly about the selection of their mobile phone. We feel that our phone says something about us, at least as much as our watch and our outfit. It is a fashion statement. We also personalize our phones with our ringing tones. We feel it must sound right to others often as much as to ourselves. Any services that help us to communicate who we are or how we want to be seen and heard, are strong candidates for Me services.

The Me attribute is how we connect to our selected communities. It is why young people learn to use SMS by sending secret notes in class, and why mobile phone messaging is so powerful in romance. In Scandinavia already women judge men by how creative their love notes are via SMS. These are shared with other female friends to evaluate the suitor. Most young couples expect the partner to send at least one personal SMS per day or else the relationship is in trouble. In the Philippines the courts had to step in and decide if a divorce via SMS was valid; they ruled that while legal, it was not recommended. The Me attribute is the most personal of the Five Ms and it is thus the most binding. Any services that address the Me, are likely to be shared. That is the underlying concept behind multimedia messaging. Early evidence from Japan's second largest mobile operator, J-Phone, suggests that the ability to share pictures on phones is perhaps even more addictive than SMS. In less than a year, J-Phone has converted over one-third of its 12 million subscribers to its picture messaging.

4.1.2 Money brings content

The fourth attribute of the Five Ms is Money, expending financial resources. Money relates to how the billing system of the mobile telecoms network can track remarkable volumes of trivial data in its charging engine. Where the internet has no built-in billing system, and the credit card industry imposes upon us minimum payments of 5 dollars or more, the mobile telecoms billing system happily tracks our every second of airtime, even if we are placing a local call. The industry talks about 'micropayments' or payments whose value is less than 1 dollar. While the fixed Internet is struggling to collect money for content — the European Internet content revenues are estimated at about 250 million euros in 2001 — the European mobile content revenues in the same year were already worth over twice that, at 590 million euros. Where most of the fixed Internet content revenues come from dubious sources (70% being adult entertainment) most of the mobile content is very mainstream, from ringing tones and logos to sports updates and stock quotes. On the fixed Internet we would hardly pay for a one-page sports update, but on the mobile internet literally millions of people already do so.

The Money attribute is much more than micropayments however. Very large value purchases can be placed on mobile phones. In Denmark they have introduced the first cash machines that allow money to be drawn via mobile phone. In Japan it is already commonplace to buy airline tickets by mobile phone, and in Finland they even accept your airport check-in via mobile phone — no more standing in lines. The Money attribute includes advertising and sponsorship. Numerous new developments are coming in this area, from forwarding discount coupons, to sponsored games and free messaging, to actually paying money to the mobile phone user for viewing ads. What the advertising industry appreciates most about the mobile phone as a media channel, is that it is the only one that allows immediate call-to-action: 'click to buy' is the catchphrase. Much more convenient than any ads on the internet, which would have to pull you to a secure website, then ask for credit card info, etc., the mobile phone can provide one-click access to placing an order. If the service is digital content — for example playing a trivia game or downloading a ringing tone — then the whole advertising, purchase, delivery, consumption and billing of the service can be completed using the same media channel. The Money attribute is the key to attracting content and to migrating content from the fixed internet to the mobile internet.

4.1.3 Talking machines

The fifth of the Five Ms is Machines: empowering gadgets, devices and automation. Machines covers machine-to-machine and man–machine interactions. This group is almost too broad to attempt to cover, as practically all modern machines are automated and most would benefit from being connected. Some of the examples are obvious, such as your car having a mapping system that accesses real-time traffic data, shows the congestion on your proposed route, and suggests alternatives.

The machines can talk to each other, such as the gas meter or electricity meter sending information to create your gas and electric bills. The utility of wireless connectivity to machines can be used in surprising ways. Intelligent copiers today can have a built-in cellular data connection and report any developing problems. The user does not have to plug the copier into a phone socket, yet the technician will learn when the device breaks down and can arrive well before the user even knows the device has broken down. In Japan the third mobile operator, KDDI, has introduced a totally automated voice-recognition and synthetic-speech-based translation system that understands spoken English, Korean and Japanese and translates it in chunks of 7 seconds at a time. It is by no means perfect, but works in most tourist-oriented situations. The machine population on 3G networks is estimated to exceed that of the human population generating new types of traffic patterns. The Machines attribute is a way to bring in cost savings from automation, and is thus the key to profits in mobile services.

Up to now, mobile services have been created by trial-and-error, and by arbitrarily porting any Internet content and services to mobile phones. The success rate has been patchy at best. The most successful services have been remarkably counter-intuitive and have had to overcome vast amounts of industry resistance. SMS text messaging is the best example of this: nobody in the industry felt it would be a mass market product used by wealthy business executives in conservative countries like the UK as much as 40 times per day. When we examine SMS by the Five Ms we can see why it is so addictive. SMS moves with us much better than e-mail (Movement); SMS can be sent on the spur of the moment where even if you had your laptop with you, you would have to power it up and log onto your server before composing an e-mail (Moment); SMS travels always directly to the intended person — the mobile phone is kept within touching distance at all times (Me); the pricing is right (Money); and we even have limited ability to use automation such as the phone numbers stored in our phonebook — thus explaining why SMS has taken so much of the traditional postcard business (Machines).

The Five Ms is a way to build compelling services. Not all needs can be met by the mobile phone, and if traditional Internet, television and print

Figure 4.1 The five Ms are dimensions and any service can be improved along each of them

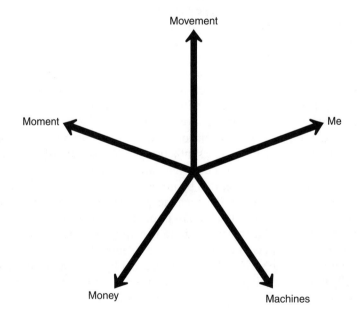

content and services are ported to the mobile Internet, disappointment will inevitably follow. WAP has provided ample proof of that. However, when needs are discovered that can be addressed with the Movement, Moment, Me, Money or Machine attributes, then compelling, even addictive, mobile services can be built.

4.2 Service management (product management)

Traditional marketing theory talked about products and product management. Since a telecoms service is a special form of product, i.e. an abstract, intangible product, both terms are used as synonyms in this book. Service management (product management) is the process of utilizing the marketing efforts of an organization to maximize the profits generated by a service. Service management typically attempts to find an optimal match between customer needs and interests — often expressed by sales and marketing — with those of the technical and engineering side of the organization, while satisfying the profit targets of accounting.

In a high technology business such as 3G it is important to have a structured and well-processed service management function. Not surprisingly, service management (product management) is one of the most central functions of a company, maintaining interfaces to all other business areas like marketing, technical department, sales or R&D. Service managers tend to be recent university graduates with engineering and business degrees. They tend to bring modern thinking and new blood to the organization. Where the employer is a mobile operator with an existing 2G network, the organization is likely to be underdeveloped and geared mostly to supplying support for connecting new customers and handling billing concerns, etc. Where the organization is a former PTT (post telecom and telegraph) and hence former monopoly, the organization may be very deep and developed, but usually with an overly technical and engineering focus. If the 3G operator is a new player, so-called greenfield operator, then all structures are new. Each of the above cases presents different issues for service management. In spite of their past history, all 3G operators will be approaching the mainstream styles of business, and will have to evolve to utilize modern marketing methods. The role of service management is central to this.

Service Management and the Service Manager is present in the entire life cycle of a product as shown in Figure 4.2.

Figure 4.2 Service development life-cycle

4.2.1 Knowing the market

We start the examination of successful service management from the angle of the services themselves. We hope to deploy successful and appropriate services. One important factor is to get to know your market place and possible customers. Market intelligence and customer segmentation were discussed earlier in this book. Often existing customers and employees provide companies with the best ideas for new services.

Too often in telecoms, research and development is technology driven. Often this will create innovative and functional services, but what also happens is that the ultimate goal is forgotten, i.e. that the service should be made for real people with real needs. Famous examples like the video recording business show that the best technology is not always the winner of the game. This was proved when the VHS standard succeeded to conquer the market vis-à-vis its technically superior competitor in Sony's Betamax. What is even more important than the technology behind it are factors like timing, the sales arguments of the product, what needs it fulfills, the right focus on the right target segment, etc.; in short: marketing issues.

4.2.2 New service ideas

Creating new service ideas starts with getting to know the market. Companies carry out their own market research, buy commercial reports to support development decisions or benchmark their competitors. In high technology industrial sectors, benchmarking is often belittled. The technical pioneer is not always in the better position for having invested often huge amounts of money in R&D. The innovative company faces the problem of early markets and has to invest even more money in convincing and market-educating promotional activities. The benchmarker, however, often enters the markets just before harvest time. In business, it is not fame that matters but fortune.

Service Development and Management 63

The history of high technology innovation is littered with failed companies who were first in the market. Few remember that before Internet Explorer and Netscape, the first web browser was Mosaic. The world's leading word processing software in 1992 was WordPerfect. At the same time the leading spreadsheet was Lotus 1–2–3 and the dominating networking software was Novell Netware. Five years later Microsoft Word, Excel and Windows NT had taken over each market. Microsoft typically benchmarks and harvests. Even on the side of the hardware most of the first personal computers have vanished such as the Sinclair and Commodore.

4.2.3 Brainstorming

One useful method for exploring new service ideas is brainstorming. In brainstorming a group of reasonably experienced people is collected together to create ideas. It is often good to have as wide a mix of know-how as possibile, and to include some people very new to the business and people who have strong exposure to other industries. The method usually works best if there is a defined moderator who understands the method well.

In brainstorming the first stage is idea generation. A popular method for idea generation is to present an inspirational introduction to creative thinking. Then each member gets a good amount of time to think of service ideas by themselves and writing all ideas, good or bad, on paper. All participants are encouraged to think of the wildest ideas, and that no practical issues are to be considered at this point, not the actual implementation, user interests, costs of delivery and revenue potential; they only have to come up with new ideas. Next the ideas are shared with the group. As each member lists and explains his or her ideas one by one, the others are not allowed to criticize or comment in any way, but are encouraged to write down *new* ideas that arise when listening to the one presenting. The only ideas that are eliminated are those that are clearly exact duplicates of what has been said. Any idea that is a significant variation of something else will be added to the list at this stage.

The idea generation phase may take several hours, a whole evening, or easily a day or more. After the group has exhausted its ability to create ideas, the brainstorming enters the second stage. At this stage the ideas are evaluated with immediate comments of duplicity of existing products — often, especially, newer people do not know that something may exist in the company — duplicity with idea entries, ideas clearly impractical in the current abilities of the company, and ideas that will definitely be rejected by management. The last category might include for example morally and legally unethical services, etc., or services that go against the company's stated strategic direction.

After this stage all remaining services are written in a list, with adequate explanations of any ideas that are not obvious or to explain distinctions between two seemingly similar concepts. Then, later, preferably soon, the idea list is ranked by some group into possibly useful ideas, unlikely ideas and impractical ideas. The possibly useful ideas listing would then be the output of the brainstorming session. Typically these ideas will still be strongly modified and eliminated. A successful brainstorming session might yield only three or four ideas that would go into preliminary development for feasibility studies, market evaluation, etc.

4.2.4 From idea to opportunity

After feasible ideas have been turned into service propositions, market opportunity and often customer acceptance needs to be evaluated. Is the target segment critically big and does it have enough spending power to buy the product? We need to keep in mind that if the service is a radically new concept, end-user interviews and focus groups may give very misleading results. The users are simply unable to imagine how their lives might change with the new services or products. For example if SMS text messaging was tested and evaluated in the early 1990s on people who had never tried it, and who probably did not own a mobile phone or had their first phones, they would probably have said that the user interface is too clumsy, that they would not use SMS. Yet actual usage has surprised all experts, forecasters and analysts. In many cases of totally new service propositions, management judgement and trial-and-error will need to be used. Next come pricing and cost analysis which will be analysed later in this book.

4.2.5 Let there be light

After the first steps in finding and scrutinizing ideas, the creation process will take place. To create a new product successfully, the different corporate interfaces have to be included. Service development is a complex process, often taking several months or more. The project team consists, in minimum, of the project manager, a marketing and/or sales person, and the future product (service) manager. Large and technically demanding service development projects can include a mixed variety of stakeholders. After all, not only crucial input is guaranteed by the different members of the project team, but also internal commitment is achieved.

There is a whole lot of questions to be answered and documented before a technology or an idea is a real product or service as listed.

Service creation questions
- what is the target segment?
- what features does the product have?
- legal restrictions;
- pricing (e.g. different pricing for different target markets such as business and consumer);
- existing competing and complementary products;
- business plan;
- marketing plan;
- appearance of the product (image, box, etc.);
- sales process — how can the customer buy the product? How is the order passed?
- production process — how the customer gets the product? How is the order processed?
- after sales process (warranties);
- support processes (web, hotline, e-mail);
- billing impacts;
- partnership issues;
- intellectual property rights, including content copyrights and any patent issues.

4.2.6 It is your own sales who knows your customer best

One of the best idea generators and messengers of new product ideas is a company's own sales force. No one has a better view of the market, competitors and customer demands than sales people. Product management usually has strong ties towards sales in support matters — where the sales force requests support information and resources from product management — but R&D should also be included in the process. The sales force should be made to understand how valuable the information they have on their customer preferences can be.

All too often reluctant sales efforts — particularly concerning B2B (business to business) products — have corrupted an otherwise good product, by forgetting to establish and communicate internal sales arguments and

Figure 4.3 Product/Service manager's responsibilities

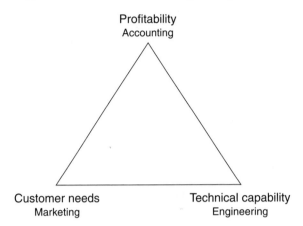

securing the commitment of the sales force to stand behind the products. Many new high technology products that are sold to businesses involve consultative selling and therefore rely even more on the good will of the sales force and an eagerness to sell the service. This is why the sales force should be given the possibility to comment on the product well ahead the launch of the product in order to avoid pitfalls. For more on the sales force and distribution channels, see the Chapters 7 and 13.

4.2.7 Caught in the middle of the triangle

Service/product managers are literally stuck in the middle. They have to satisfy the needs of the customer, what the customer hopes and expects. This could be summarized as a marketing need. They also have to deliver something that the network itself can sustain without problems: a clear engineering need. Finally the whole service proposition has to be profitable, satisfying an accounting need. This whole issue could be described as a triangle, where the three corners are customer, network and profit.

A typical conflict can arise when the service portfolio is found to have an existing service with a bug, and an upcoming service has an urgent deployment schedule. The service creation people or the service manager(s) involved will have to decide where to put the time and resources: do they satisfy the existing service repairing need or the new service creation need.

The service manager is stuck in the middle, trying to satisfy conflicting interests. Due to this inherent conflict, the service managers are likely to be among the most stressed professionals within a modern telecoms company. Management should recognize this inherent level of stress and ensure that properly trained and supported experts are given this role, and also that the performance measurements and compensation of service managers is in line with the challenges of the job.

4.3 The launch

The launch phase is one of the most critical moments of the product management process. Of course activities will differ depending on whether the product is the first of its kind on the market or if a company enters a well established market and is fighting for market share or a certain niche. The proverb 'you never get a second chance to make a first impression' applies to all activities with a new service launch but is particularly true when thinking about press relations. Some of the relevant issues at launch relate to service adoption and whether the launch is directed at early adopters, etc. For more see the Chapter 11.

During the launch, the internal marketing activities are often as important as external marketing efforts. It is your own company and sales force who have to be convinced first that the service is viable and worth selling. In this light also, external marketing activities such as advertisements help increase your own staff's faith in the service. Of course, this applies to a big extent to the B2B sector.

Product launches are good business engines that serve to illustrate that the company is viable, vibrant in the market place and has R&D power. This will be increasingly important in the mobile Internet area in attracting partners. Big product launches are good occasions to boost press relations. Do not miss a good occasion to motivate employees by organizing a festive launch event.

4.3.1 Tariffing, cost and profit

Tariffing, costs and profitability are often the source of the biggest headaches for product managers/service managers. The service manager's most important goal is to deliver a commercially viable and profitable service. However, all the cost factors that partly contribute to a product's price often cannot be identified. Many costs are hidden or difficult to calculate. Fixed costs may be assigned arbitratrily by history on organizational structure rather than any practical reasons. Tariffing, which usually is the most important decision, is also often almost arbitrary and can be done with little scientific process. Tariffing is such a significant issue that it is discussed in its own chapter (Chapter 14).

It has clearly been shown in many parts of this book, that pricing in 3G has to be done in a new and purposeful manner: we have to detect where the pain threshold is and locate the services's price near to it, but below the pain threshold in order not to kill the business in the first place. Operators will need to learn to identify tariffing points and price elasticities.

4.4 Killing a service

Of course it seems funny to speak about killing services in a business that is presently giving birth to the first services in the market. However, terminating a service or product is a natural occurence in the service/product life cycle and service management needs to understand it. In the 3G world, the service life cycles will be much shorter than what have been the cycles in traditional telecoms. This makes killing a service more important to 3G than what it has been in the past. Furthermore, with vastly larger amounts of services in 3G and the mobile internet, and shorter product life cycles, the termination occurences will happen much more frequently.

Why would you want to terminate a service. There are several reasons to terminate a product.

Reasons to terminate a product
- image (the product does not fit in the portfolio any more or does severe harm to the company's image);
- selling the service exclusively to another service provider altogether is more profitable than selling it on its own;
- no profitability (e.g. if the service is tying up too many resources or is not profitable in the long run);
- lack of customers (the maintenance of the product is more expensive than the revenue generated);
- cannibalization (a product is eating the profit perspectives of another — this only ties up resources and is to be avoided in the long run);
- service management load (the product causes too much work).

4.5 To finish with service creation

This chapter has looked at service development and management. These are vital for success in 2.5G and 3G. Initially many services will be directly ported from other media and applications. Soon true mobile services will be developed with methods like the Five Ms. Beyond that, real creativity is needed to discover innovative services and techniques such as brainstorming will be needed.

This chapter also gave an overview about service management. Of course all of 3G is about services, and 3G marketing is essentially all about service management on some level, so most of this book applies to a service manager's various areas of responsibility, such as customer understanding, tariffing, promotion, sales support, etc. It is important to make sure that early on in 3G creativity will flourish. When guiding creative service creation staff, it is good to keep in mind what Henry Ford said: 'I am looking for a lot of men who have an infinite capacity to not know what can't be done'.

Man is an animal that makes bargains; no other animal does this.
— Adam Smith

5

Partnership Management
Teaching an Old Dog New Tricks

Where product management/service management was living through the stages of establishing itself in telecoms during the 1990s, partnership management will be going through a similar evolution during the first decade after 2000. There are strong parallels when examining the role of a partnership manager and that of a product manager/service manager. Both have to try to deliver services that customers want, that the network can deliver, and that can do that at a profit. Both jobs, therefore, are at the epicentre of the conflicting needs of each of those three dimensions.

3G Marketing: Communities and Strategic Partnerships   Tomi T. Ahonen, Timo Kasper and Sara Melkko
© 2004 John Wiley & Sons, Ltd   ISBN: 0-470-85100-7

Figure 5.1 Partnership manager's responsibilities (theory by Tomi T Ahonen and Marc Hronec)

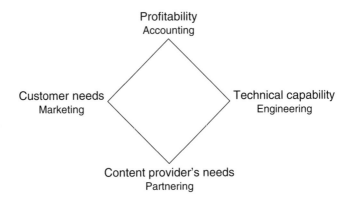

The partnership manager, however, has a further dimension: to satisfy the need of the content provider (and/or application developer, content aggregator, IT integrator or other such third party). Where the product manager/service manager previously had to carve out his/her position amidst established departments and units *within* the company, the partnership manager has an even harder job of convincing the organization to deal with *outside* organizations. That job will not be easy and the partnership manager is likely to become the most stressful job in all of telecoms. Its successful completion however, is quite literally the key to profits in new mobile services whether 3G, 2.5G, W-LAN, 4G or any other wireless technology (Figure 5.1).

Essentially everything about service creation and service management in 3G applies also to partnership management, therefore in this chapter we will not re-examine those issues. We will proceed to the specific issues of the new, fourth dimension, partnering.

5.1 What is partnering?

'Partnering' is a term very often used in business today. Every provider seems to be saying that it wants to deepen its relationship with its customers and move from a customer relationship to a partnership. Most vendors and suppliers insist on wanting to develop a deeper relationship by becoming a partner. It is important to understand that, fundamentally, partnering is a deeper commitment than mere sales. That concept helps to identify our definition for this chapter and for the whole book. Where a customer is purchasing goods or services, that is sales. Partnership needs to be more than that. It is a clear differentiating factor. If the mobile operator sells, for example, premium SMS services at a certain bulk cost per message, and makes that offer to a given content provider, then that is not partnering. For partnering to happen something else must take place as well.

5.1.1 Flavours of partnering

Moving past a customer relationship, partnering can take very many forms. One traditional way to partner is for the two parties to barter or exchange their respective offerings, in effect paying for each others' goods and services in kind. A barber gets vegetables from a greengrocer who in turn gets his haircuts at the barber, so an informal partnership has emerged. This is a frequent form of benefit that individual partners can give their counterparts, provided of course that the recepient has real need for that service. A purely financial reason to do this is that for the party providing the goods or service the value is that of its net value, while to the receiving party the value is that of the gross value. A more corporate and modern variant is an exchange of equity between partners.

A second form of partnering is that of exchanging confidential information. It may be the pooling of news gathering, for example with news organizations, or it could go much deeper and involve giving access to a partner into internal databases and intranets, etc. In many cases this can be critical in enabling a partner to be better able to serve one another. It, also of course, exposes the parties to risk of information abuse and leaks via the partners, in particular due to the rapid development of IT infrastructure. This is usually the level when conservative elements within a telecoms operator — or in many cases also the equipment vendor — will become alarmed. They can appreciate the benefits of partnering by *gaining access* to resources and information of the *other side*, but become increasingly worried about *revealing internal proprietary information* to the so-called partner. For partnering to work at this level, the sharing will need to be equivalent, at least with regard to the area of business where partnering is taking place.

Partnering can take the form of cooperating to serve common customers. At a minimalist level this can be marketing and promotional materials that mention the partner, with co-branded logos, etc. It can also mean joint visits to common customers, especially larger corporate customers. At deeper levels of partnering, the customer contact and sales can be expanded fully to include each others' offerings. I sell your services and you sell mine. In such cases the partnership needs to determine sales commissions and/or leads that relate to such sales.

Partnering can go ever deeper into various cooperative projects. The significant key is that some common goal should exist, for the reader to recognize that partnering can include different types of involvement and cooperation, and partnerships as such can evolve.

5.1.2 Who are the prospective partners?

The main candidates for partnering in new wireless services include the mobile operator, the portal operator, a content aggregator, the independent sales channel, the application developer, the system integrator, the content provider, the network infrastructure vendor, and the handset supplier. Each of these players has his own interests and will of course pursue any potential partnership with a profit motive or similar gain in mind. As is typical in 3G, in many cases any prospective partnership could be conducted with more than one of the alternative partners.

Mobile operators are not accustomed to working with dozens or, potentially, hundreds of small content providers and application developers. The content providers and application developers tend to approach the new service opportunity with perhaps an over-optimistic view of what can be done 'on a stand-alone PC' or other such simple service platform. These

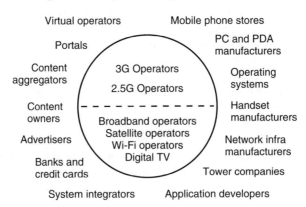

Figure 5.2 Players looking for a role in mobile

Partnership Management

may be acceptable when users number in the hundreds or thousands, on a service that runs with an ISP (Internet service provider) for example. However, the requirements of mobile telecoms operators, with easily 10 million subscribers, are very different, and among the first concerns of product development with major operators are reliability and scalability. Will the system stand a simultaneous request by 500 000 users? Will the new solution allow fraud into the network? Is it possible for the new system to crash the whole network? These may seem like paranoid concerns to a young game developer with the coolest idea on earth, but a cellular carrier cannot take the risk of its network going down for even a few minutes. The traffic loss and bad publicity is not worth it.

On the other hand, mobile operators are not known for innovation and fast deployment of new services. They are used to product development cycles of 18 months, where their equipment vendors brought roadmaps of the next evolutions in technology 3 years into the future. That is a far cry from the Internet product development cycle, where 90 days is commonplace, from idea to commercial launch. Here cellular carriers have to respond to market needs and speed up their development. Several partial solutions exist from specialized 'middleware' service creation platforms that allow fast implementation of solutions, which then connect in a standard way to the various management systems, CRM (customer relationship management) systems, billing and provisioning, etc. Another way is to have in-house test networks, for example in a given city or for a given customer segment, where services can be deployed on a smaller scale to allow trialling and learning.

5.2 Operators are new to this game

When examining partnerships in new wireless services, a vast array of potential pairings and combinations are possible. The most likely player to be involved will always be the mobile operator. It is the only one able to allow access to any given user on any given device to the cellular network. Where the sheer volume of partnership prospects may overwhelm the ability of the mobile operator to handle all requests, and thus perhaps he may need to delegate the role in many cases to content aggregators, independent portals and even MVNOs (mobile virtual network operators), the decision to forego a role in any given partnership is with the mobile operator.

On the most basic level, most executives at mobile operators do understand the need to partner. They recognize that they cannot create all possible content and applications for thousands of services all by themselves. In the more practical day-to-day operations, the mobile operator is new to the challenges of managing multiple partnerships. Recently many CEOs have been calling out for developer communities and promises of developing ecosystems. There is a wide gulf between the lofty ideals promoted by some forward-thinking CEOs and the actual day-to-day partnership management conducted by typical mobile operators.

We must keep in mind that the whole mobile Internet came into existence only in 1998, and the earliest knowledge about how the new mobile-services world functioned started to emerge in 1999 and 2000. There has been almost no time to consolidate knowledge and most advances have been through trial and error. Mobile operators have initially been cumbersome and made blunders in their dealings with potential partners. This was not due to malice, but simply to inexperience. The other parties, who may be much more familiar with partnering in other industries and with other communication media, should be tolerant of mobile operators as they follow the steep learning curve to become contributing and caring partners for the long run.

5.2.1 Culture shock

Often when members of a mobile operating company or telecoms equipment maker meet any other players in the prospective partnership arrangement, a strong sense of culture shock emerges. Many of the staff of mobile operators and equipment vendors are telecoms engineers with a long history in the industry. They may have started in the industry when telecoms was still regulated and even a monopoly. If not, then it is very likely that most of their colleagues and managers are that kind of traditional telecoms engineer.

It should be borne in mind that telecoms as an engineering industry is among the most conservative purely for long-range planning and implementation reasons. Any new device connected to the telecommunications network will be connected to all others, and the global telecommunications network is the biggest interconnected system that man has ever created. It is backwards compatible with generation upon generation of older technologies and proprietary national standards, etc. Telecoms engineers have evolved therefore to be careful and test a lot, and standardize a lot, before implementing radical new technologies. Nobody wants to go down in history as having caused major network crashes involving millions.

Therefore, telecoms engineers are much more traditional and conservative than those often thought of as their close siblings, namely IT and computer

Partnership Management

engineers. With the advent of the personal computer, computer engineering took off quickly and started to attract creative people and companies with rapid development cycles. There was no need to standardize everything as long as given components, software and systems worked with some of the computers. The IT industry is full of university drop outs who never finished their degrees as they turned to making money out of programming. These engineers are at the utmost opposite end to telecoms engineers, more creative and less formalized in their patterns than any other engineers.

As these two mindsets meet in a partnership discussion, almost all possible communication conflicts emerge. A basic philosophy of distrust and lack of respect tends to exist, where telecoms engineers feel that their partners are dangerous and that they totally fail to understand the needs of the telecoms network. IT oriented engineers tend to think that telecoms engineers take forever to decide anything and are about to become obsolete as did the dinosaurs. The very languages of telecoms and IT differ, with both industries developing competing synonyms for most similar technical concepts. Whole dictionaries exist to translate telecoms terminology to IT terminology and vice versa. The culture differences can usually be seen with dress code. Telecoms engineers tend to dress in suits and ties while, in particular, the

Figure 5.3 Revenue Spilt along the Value Chain

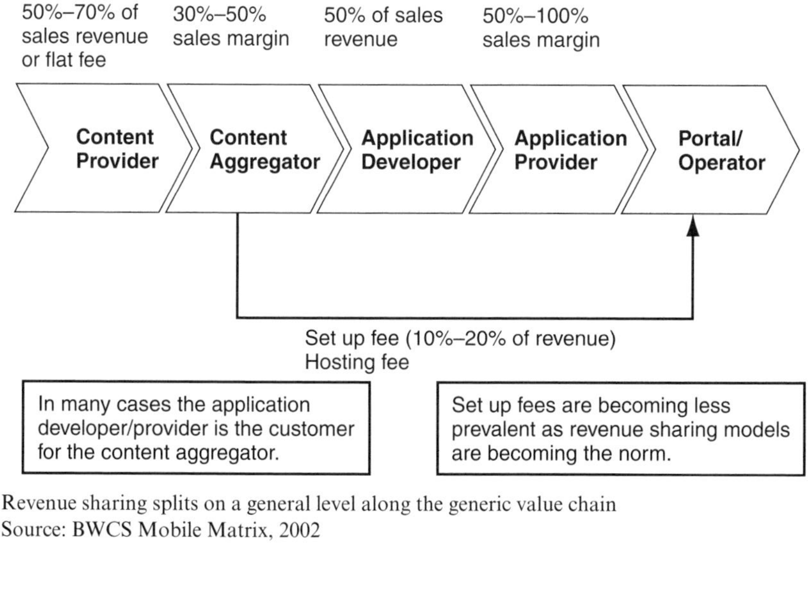

50%–70% of sales revenue or flat fee	30%–50% sales margin	50% of sales revenue	50%–100% sales margin	
Content Provider	Content Aggregator	Application Developer	Application Provider	Portal/ Operator

Set up fee (10%–20% of revenue)
Hosting fee

In many cases the application developer/provider is the customer for the content aggregator.	Set up fees are becoming less prevalent as revenue sharing models are becoming the norm.

Revenue sharing splits on a general level along the generic value chain
Source: BWCS Mobile Matrix, 2002

smaller-application development staff tend to be in proverbial ponytails with no discernable dress code whatsoever.

The activities in partnership meetings will differ radically. As telecoms tends to move slowly, but the requirements are strongly dependent on standards, most telecoms engineers prefer to use pre-prepared Powerpoint slides of sets that may easily remain the same for years on end. This frustrates any creative person who wants to do something new. The most creative application developers tend to want to work from blank sheets and whiteboards and to draw up new ideas and ways for services. This impulsive method of working is very alarming to most telecoms engineers who may feel that there is no credibility to such way of doing technical work.

The same is true of contacts, e-mail etiquette and follow-ups from meetings, making decisions, etc. A telecoms engineer tends to take his time to come up with a decision, but also tends to be diligent in replying to e-mails. Many of the younger generation of applications developers can be very impulsive in their decisions and can change them many times over as newer and better ideas emerge. They are also notorious for not replying to e-mails. All of these differences in culture are major causes for communication failure and can often cause rifts and create mistrust between the parties. The reader should be mindful of these differences and, whenever meeting someone from a new partner prospect, should start off early by discussing these issues and explaining how and why the culture tends to be as it is in their company. Understanding is the very first step in getting over the conflicts of culture.

5.3 Revenue sharing

The principle of revenue sharing is rather simple. Two or more partners bring in their own 'value-add' to the service proposition, and the partners share in the revenues from the resulting new business. When we examine this idea in more detail, numerous issues emerge. It is important to note that at the writing of this book, in the winter of 2003–2004, the revenue sharing mechanisms were not yet set in stone. The whole revenue sharing proposition was still undergoing considerable growth and evolution. Initially most operators offered revenue-sharing deals of roughly 50:50 for content revenues (and no sharing of traffic revenues). It was not until the introduction of NTT DoCoMo's I-Mode to Europe and similar propositions in the USA, that the whole industry was forced to examine what would be fair for the mobile operator to keep in the revenue sharing proposition.

Partnership Management

While I-Mode's Japanese model of keeping 9% of the content revenue and returning the remining 91% to the content provider has been practical in Japan, the exact same model has not been introduced elsewhere. Typically, however, the current content-revenue sharing ratio is starting to approach a level of around 12–25% to the mobile operator and the rest returned to the content provider. It is also important to bear in mind that mobile operators as a rule do not share data transmission revenues.

5.3.1 What kind of revenue (and/or cost) sharing options?

The range of revenue sharing options is broad. While an application developer, portal, IT integrator, and/or content aggregator may also be part of the partnering equation, for the sake of simplicity let us assume only two players for a given service partnership: the mobile operator and the content provider. At one extreme the operator could purchase the full rights to cellular delivery of the content. These could be exclusive (no other mobile operator may offer the content) or non-exclusive. At the other extreme the operator may want explicitly to have no right and no affiliation with the content. An example of this could be adult entertainment.

Between those extremes, the operator and content provider could come up with a wide range of revenue sharing arrangements, such as a one-time payment with limited rights, a limited amount of use, or limited amount of time that the content may be offered; revenue sharing based on billing of the service, such as for streaming music; or a usage-based revenue share, with a split based on time used with the content such as a game; the number of clicks or page views, etc.

The venture may have particular risks on one side or the other, and part of the revenue-sharing benefit or burden may be shifted to cover the risk. For example, a new pop-culture-based cartoon character, for example, relating to a movie that has not yet been released, might have a certain minimum usage level defined before it could generate any revenue sharing to the content owner, to offset the development investment that the operator has to take just to enable the content.

For various services there can be ceilings, i.e. maximums, that can be paid and floors, i.e. minimums, that will be paid to whom and under what conditions. If the concept is strong enough, the operator and content provider could set up a joint venture and split its profits, as is happening with m-banking and mobile advertising in many markets.

5.3.2 What level of revenue sharing

When examining an industry soon to be worth a trillion dollars, certainly the guidelines on what to expect in revenue sharing are at least a 'billion dollar question' for any given industry. The different expectations of what any given partner might gain out of revenue sharing will vary greatly from one industry to another, and among players within an industry. A thorough examination of the existing and evolving value chains in any given (non-telecoms) industry must be examined as a starting point. To that end, the analysis needs to include the relative value propositions being brought to the new service idea, the relative strengths of the brands involved, and the extent of possible substitution both in terms of content (and/or application) and delivery channels.

Let us explore two examples. Delivering news content from CNN and delivering game score summaries from the local sports league. With CNN the news content is collected at considerable cost from exclusive stories from the news staff flown around the world to cover major stories. CNN brings a level of credibility and is one of the largest organizations with its own journalists and electronic media technicians anywhere. A mobile operator based, for example, in France would find it almost impossible to attempt to staff a reasonable worldwide journalist staff to compete with CNN's coverage. Of course the mobile operator could purchase rights to create stories from news agencies like Reuters and AFP, but then the mobile operator would be relying on another third party content provider.

In the case of CNN content, for the most part, the content would be unique or a comparable audio-visual current news offering could be found from CNN's close competitors like Bloomberg, CNBC, BBC, etc. As no viable solution by the mobile operator exists, and clearly judging by the number of 24-hour news offerings on cable and satellite TV, a vast appetite exists for current news coverage, the mobile operator would need to partner with a major news operator. Also, as the mobile operator would have no significant option to do it by itself, the operator would have a relatively weak negotiating position from which to insist on a major share of the revenue sharing arrangement.

Now let us consider another part of the news, the scoring updates of the local sports leagues, be they football, basketball, ice hockey, cricket, etc. These leagues have it in their interests to publicise the scores and results, and will be offering that amount of information to practically any outlet that is willing to cover their sport. While not allowing access to personality

Figure 5.4 Revenue share returned to content owner by region, showing the range of revenue sharing deals offered by operators in the region (Nokia November 2003)

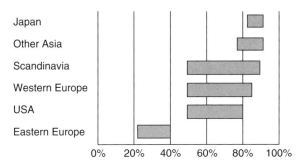

photos and game video images — these tend to be sold via exclusive deals to various media networks — the scores themselves are as near to generic news feed as is conceivable. In this case any sports service which promises to deliver the game scores covers very little of what the mobile operator could do, and such news offering would be of relatively little value to the mobile operator. Hence the mobile operator would insist on his least favourable terms for such a content provider, and equally, that the operator's own small news team would be likely to offer the same content.

With such a comparison we can see that the two would behave very differently as mobile services, and the operator and content owner could expect different proportions of the revenue share, based on the added utility, availability of alternatives, and strength of the relative brands. Each industry will need to be examined and the role of mobile services needs to be explored. Several industry-specific reports exist that examine the potential role of the players in the emerging new mobile service market. A comprehensive revenue sharing report has also been released by Tomi Ahonen Consulting, which examines the overall trends of revenue sharing and partnering during 2002.

5.4 Main factors influencing split in revenue share

There are several main factors which have an influence on determining the ratio in revenue sharing. These include exclusivity, the value chain, on-screen location, the relative strengths of the brands involved, location information, billing information, user information, etc.

Exclusivity is the first determinant. If the service needs a mobile operator for the service to function, there is a considerable amount of exclusivity as most other operators, such as cable TV, fixed telecoms and ISPs, are not viable options. One example could be services delivered in real time to a moving car. Currently no digital TV or W-LAN (Wi-Fi/802.11), etc., solution can deliver real time services to moving cars. In addition to exclusive services, some of the end-user benefits can be absolute competitive advantages when compared to other technologies. Those will be covered in Chapter 11.

5.4.1 Exclusivity

Furthermore, there can be exclusivity among mobile operators, depending for example on a given technical standard, the availability of mobile phones and other terminals, geographic cellular coverage, etc. The fewer the viable options, the more the mobile operator could expect to get as its share of revenue sharing. The same is true of the content. Madonna's music is Madonna's music and no copy will serve. Madonna's agent and record label get to make the final determination of how her music and brand is deployed in the mobile Internet space, and as there is no alternative, it is either her nor nothing. She holds a very strong negotiating position.

5.4.2 Value chain

The existing value chain or value chains are a significant determinant. In many industries the delivery channel has introduced digital delivery and these are likely to have the mobile Internet delivery option and its portion of the overall value chain relatively well defined. The mobile operator's share would be similar to what has already been introduced. If the industry itself is still in the early phases of conversion to digital delivery, then there is much more latitude for determining the mobile operator's share. The extent of the value-add by both parties is of course also important. For the mobile operator, the main elements of direct value-adds are location information, access to user data, the micropayment mechanism, and the ability to connect to direct communication systems, mostly voice and messaging. The more operator value-adds used in building the service, the bigger the operator's share can become. Equally the more value-add elements the partner brings into the service, the more that partner can expect to get in its share.

Partnership Management

Figure 5.5 Music Market's Value Chain.

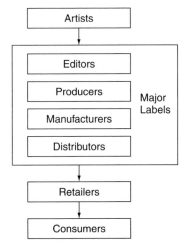

Source: IDATE 2001. As more music goes online and is downloaded electronically as 'bits' of data the existing Music Value Chain will evolve and become more efficient and streamlined. The role of retailers and distributors will change and some artists will develop direct channels to their fans. This will only happen however when Digital Rights Management that protects Artists' copyright is introduced.

5.2.3 On-screen location

On-screen location is another key factor. The ideal position for accessing a mobile service is right at the welcoming screen on the mobile phone. With very little total available space on the screen of the mobile phone, and with thousands of potential services vying for that location, the operator is in a strong position if the service is set to be close to the top of the screen. The role of the portal will be significant, as well as how it is configured and modified in determining which content gets what placement. The service can also be built to be behind one or two keystrokes, which would have a similar effect. The higher the service is placed near the top of the opening screen, the more the operator could expect to get in its share of revenues.

5.2.4. Brand strength

Finally, the relative strengths of the brands are a factor. Brands will need to be considered always in the context of the given service and its intended target audience. There are numerous racing car games, but Formula One would have strong brand awareness that would bring more loyalty and users to its game than to any generic racing game that might be deployed. That brand would probably find strong awareness in Europe, Asia, Australia, Latin America and Africa, but not in North America, where Champ Cars, Indy Cars, NASCAR racing, etc., would be much stronger brands than Formula One.

5.2.5 Location information

Location information is a powerful determinant for some services. If location information is required for the service, such as finding the nearest Italian restaurant or providing a location-aware map, then the mobile operator is by far the most efficient way to provide this information. Other alternatives include the user typing in the street address or postal zip code for the location, or extracting the information from a GPS (global positioning system) but these are very cumbersome or costly, in particular with regard to the extreme time-sensitive nature of mobile services. If we sit at our computer it may not be a major issue to type in some information, but when rushing about town and trying to get a mobile service to work for us, the time and convenience is very important.

5.2.6 Charging/billing information

The charging engine is one of the competitive advantages of mobile operators. We will explore its power more in the tariffing chapter (Chapter 14). We have already discussed some of the incredible user behaviour data that can be extracted when we covered segmentation. If the partner wants to know about user behaviour, preferences, use of services, etc., then the telecom's charging and billing system provides a goldmine of information, unavailable in any other industry. This can be of remarkable benefit for most services and is something that the operator can use as leverage in setting up partnership arrangements and revenue sharing.

5.3 Rules of thumb

Three simple rules of thumb can be taken from the digitally converging world. In the credit card industry, the credit card companies tend to have commissions of between 2% and 4%. If the mobile operator offers no added benefit whatsoever from location information, subscriber data, urgency/timeliness, and only provides the billing convenience and takes a risk replacing a credit card company, it is reasonable to assume that the operator would take at least the same amount as the credit card company. As most credit card companies have minimum payments, if the operator wants to offer micropayment options (payments with a value of less than 1 dollar), then the operator could be expected to keep more than the 2–4% that a credit card company does.

The other extreme is splitting in half. While this may seem quite extreme, in many emerging digital services that need the operator's know-how and systems, for example developing a multi-user, real-time, network game, it is fair to assume that the operator gets a considerable part of the revenue due to its considerable involvement. For most combined service developments, a 50:50 split would seem the other extreme. In this case the operator provides location information, customer data, and billing services to the partnership, as well as its branding and location on its portal. Also the operator would take an active role in the system development or integration involved. It should be noted that while early on in the mobile internet a 50:50 split of revenues could be 'forced' upon some application developers and content providers eager to get into the mobile services arena, the evolution towards more realistic market conditions has brought that extreme closer to 25% at the other extreme. So as a second rule of thumb, we could use 25% as the 'most operator-friendly' sharing proposition. It is the opinion of the authors that operators will soon be unable to squeeze more than that even under the most ideal conditions. So the practical range of revenue sharing should be somewhere between 4% and 25%, depending on the various conditions discussed above.

A third rule of thumb is the I-Mode example. DoCoMo's I-Mode service is widely quoted as taking a 9% revenue share cut from the value of the content or subscription to the official I-Mode sites. Outside of Japan that has been often converted to approximately 12% or so. If no better initial benchmark exists, this can be a good starting point. A fair approach could be that starting from 12% the two parties could examine what are the relative merits that each partner brings to the partnership, should the revenue share be more or less than with I-Mode. From that the relative extremes could be about 4% as the absolute minimum and 25% as the absolute maximum to go to the operator. The reader should, furthermore, keep in mind that the revenue sharing arrangements are still evolving during 2004 and these rules of thumb are not likely to remain fixed.

Figure 5.6 An example of revenue flows for the UMTS operator and service provider

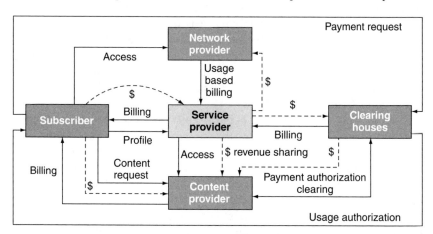

Market rules will also enter into play. The content providers must be active in seeking opinions from all players in their market — not only the actual mobile operators but also any independent portal services, content aggregators and MVNOs. The content providers must use this information as leverage when playing an offer from one operator against the other to secure the best deal. Similarly, the network operators must remember that in their portfolio of thousands of services, there are rarely unique service offerings, and they too should be in contact with several candidates to find the preferred partners.

5.4 Contract management

One of the new aspects of partnership management is the mechanical exchange and mutual acceptance of partnership agreements. This typically involves some generic contract 'boilerplates' by the mobile operator (and/or portal, content aggregator, MVNO, network platform developer or handset manufacturer). The interaction between offer of partnership, construction of relevant clauses into the draft contract and the mutual acceptance of the contract terms with the departments and usually also an attorney by both parties, is a process that early on can take weeks, even months. This process must be streamlined and automated as much as possible. Already some technical solutions are being developed to manage the process of going

Partnership Management

through the 'contract ping-pong' and the acceptance of various requests and terms that typically a mobile operator would need to cover the multitude of departments that would be involved. These include service creation, service management, accounting, network dimensioning, billing, etc.

A goal for the 2.5G or 3G operator must be the ability to handle the full acceptance cycle in 48 hours or less. That means that almost all interaction must be automated, with all basic information available on-line and automated processes to flag problem cases that may need human intervention. The approval process itself should migrate toward one that is totally automated, and for those who wish to enter into basic level contracts with the operator (or virtual operator, independent portal or content aggregator), approval should not require human managers to accept.

5.4.1 Keys to success

The chapter on service creation (Chapter 4) showed how any content can be enriched with the Five M's when brought to the mobile Internet. However, even the best service or product may fail in the marketplace. More typical of both the IT and telecoms industries is that success comes suddenly from unanticipated sources. A good example is DoCoMo's I-Mode service, which at launch expected to have its success coming from businessmen gambling on sports. Most of the revenues in fact come from young people playing games or downloading cartoon characters and other such youth-oriented entertainment. This has obviously been a surprise in the Japanese market.

Similarly, in Europe, the counter-intuitive SMS (short message service) text messaging service was not seen early on even as a viable business by most European operators, until some of the early data started to filter in from early adopter countries such as Finland. Now every mobile operator around the world offers SMS text messaging and even the laggard USA has caught the fever with SMS traffic on their networks exceeding 1 billion messages per month late in 2002.

New success can mimic success in the entertainment, gaming and pop culture areas. Content built around TV shows with large audiences, such as *Who Wants to Be a Millionaire, Survivor Island, Pop Idol,* and *Weakest Link,* seem likely candidates to succeed in the areas of mobile Internet content. The same would go for Harry Potter and *Lord of the Rings* movie tie-ins, as well as anything built around the latest pop music stars. The content can take on a multitude of formats, from direct content (songs, video clips, foreign language versions, books, etc.) to news, gossip, backgrounders and analysis, and on to games, cartoons, merchandising, etc. Remember also that successful content can appear seemingly out of nowhere and grow fast to remarkable popularity. Many a popular phenomenon can also disappear fast, as the tamagochi phenomenon has proved. Telecoms operators are accustomed to building services that have a lifespan of decades; much popular content might have a lifespan measured in months or even less.

5.4.2 Partnering for profit

While a partnership may sometimes be created for purposes such as filling in a perceived hole in a service offering, or to act as a loss-leader, most partnerships should yield services that are profitable. Telecoms operators are not known for understanding the profitability of their products particularly well. Here is one area where the partners need to assist the partnership by keeping the profitability of the partnership venture in focus. Profitability needs to be calculated for every separate service, every target segment, as well as any given service bundle. When considering a large new portfolio of new mobile services, a mobile operator has to give its controllers, partnership managers and service/product managers clear guidelines on how to determine profitability, so that the profitability calculations and measurements for the whole product range are compatible and comparable. Often this includes a management indication of network costs including both operating costs and capital expense. With most telecoms operators this type of information is held to be extremely sensitive and only available to a tiny part of senior management. In handling the profitability of a portfolio of mobile services, it simply is not acceptable to hide the real costs from those whose job it is to manage the services and oversee their sales.

Lastly, successful partnership management comes down to people. A few years ago no mobile operator staffed teams of partnership managers. Now all will need to institute this new function into their organizations and to hire teams, even departments, of partnership managers. The commercial success of the advanced services will depend more on these new partnership managers than on any other people or resources that are to be deployed. The skills that a successful partnership manager will have are those of an exceptionally astute and successful product manager/service manager. Mobile operators could well institute the job description as a logical career path and growth for the best of their staff of product managers/service managers. Of course the partnership managers will need skills in engineering, marketing and accounting just like product managers/service managers. The additional skill set could be summarized as being that of attorneys, and no doubt most partnership management teams or departments will include a few trained attorneys as well.

5.5 Parting with partnering

More new mobile services will be made with partners than alone by any of the major players, including the mobile operators. For a partnership to work, both (all) parties have to find real benefits on a long-term sustainable basis. Revenue sharing will be a key ingredient to such partnerships and mobile operators will be on a steep learning curve to achieve the ability to manage such partnerships. Some partnerships will end up not working and others will run a natural course and end. Most partnerships will have their challenges and will require work and effort. When the first problems arise the easiest path is to abandon the partnership and seek a new partner.

While it may be difficult, and at times frustrating, to work with new partners, it is best not to abandon new relationships early, but to give them extra effort in the early stages of new technologies. At the least the people involved will learn from trying to make the partnership work. And in the best cases going together through the difficult times will make partnerships stronger. Building a partnership in difficult times will require a certain degree of optimism. A guiding light can be obtained from what John D. Rockefeller said: 'I always tried to turn every disaster into an opportunity'.

The Americans have need of the telephone, but we do not. We have plenty of messenger boys.
— Sir William Preece
Chief engineer of the British Post Office, 1876

6

Terminals
A Swiss Army Knife or a Corkscrew

This chapter delves into the diversity of end-user devices. Currently most handsets are mobile phones. In the 3G future increasingly the 3G devices can be anything from the traditional cellular phone to a PDA or computing device to a household gadget, a gaming device, our car, etc. It is important to recognize that a multitude of new classes of 3G devices will emerge, including such exotic items as 3G clothing and robotic 3G cameras, as well

---

3G Marketing: Communities and Strategic Partnerships Tomi T. Ahonen, Timo Kasper and Sara Melkko
© 2004 John Wiley & Sons, Ltd   ISBN: 0-470-85100-7

as mobile telecoms technology being built into various other devices ranging from the photocopier to street lamps. In this chapter we discuss how the usability of services should be taken into account when selecting the devices that the services will be used with. Those decisions are crucial in buildling the overall marketing plan.

6.1 How our gadgets evolve

Remeber the late Sixties? In the late Sixties, stereophonic radio broadcasts and recordings became popular under a common denominator of hi-fi (high fidelity). At that time there were practically only two viable mass media for storing and reproducing audio experiences, either records ('long play' record albums called LPs, and singles) or open-reel magnetic tape-recorders. During the early 1970s the c-cassette emerged and soon became the preferred choice as a mass market recording medium. The hi-fi equipment was expensive and manufacturers did not have standardized interfaces to support a combination of these different components of a hi-fi-system.[†]

The point is that in the 1970s we had to have several different components to form a hi-fi system and often the differing parts did not work very well together. The audio entertainment industry provided us with three incompatible technologies: one for listening to radio broadcasts (the radio), another for reproducing pre-recorded music (record players) and a third one furnishing us with an option of capturing either of the previous ones (tape recorder).

Later, equipment manufacturers started to pack all functions into one unit, as the first 'combo'-sets were introduced. However, the combined sets had serious drawbacks, if you had a malfunction in your c-cassette player, you had to take your combined device including its built-in record player and radio to the repair shop. Another problem was that it did not allow you to 'upgrade' your equipment gradually, you had to buy a whole new set once you got annoyed with a low performance of one integrated component. This could provide a serious and continuous cash-flow problem to those who were serious about their music, as each of the authors of this book have always been.

[†]DIN (Deutsche Industrie Norm)-based connectors were probably the most common way to connect two pieces of equipment from two different manufacturers at the time.

The home entertainment industry was busy, however, inventing new delivery and storage media for music and other entertainment. The industry introduced several new technologies like VCRs, music CDs, the MiniDisc, MP3-players and DVDs within a relatively short time frame. This made it imperative to have some standard interfaces to enable connectivity of the equipment for the newer technologies. All-in-one systems could not accommodate the new technologies and the whole set became obsolete. This swung the pendulum back towards separate elements in the music — or now the entertainment — system. So, at the end of the day, independent components once more became better value for money and more popular.

6.1.1 Convergence

The convergence of the media-related industries has been an ongoing transformation since the 1990s. The convergence has engulfed print media (newspapers, magazines, etc.), electronic document media (any document in writing stored and distributed in electronic form) as well as multimedia (including sound and vision/moving picture). In other words, the limitations of sticking to traditional media have been abandoned; newspaper publishers maintain very popular Internet portals, and, for example, the famous sci-fi/horror author Stephen King has published a book on the Internet.

What does 3G contribute to this phenomenon then? We are in a similar situation with high capacity wireless communication as we were when the optical discs (regular compact discs and (re-) writable CD-ROMs) were introduced in the entertainment and computer businesses. The device and the basic technology itself are used in an increasing number of applications. 3G will introduce to our everyday life changes almost as dramatic as the microchip has done so far. You can find microchips in almost any device nowadays, not just in computers. They exist in children's toys, washing machines and other household appliances, in cars, wristwatches, and multi-functional bicycle speedometers, just to name a few.

The capacity of 3G will enable services of exponential variety as compared to other wireless technologies. To call the terminals 'mobile phones' is already an understatement, even the most inexpensive units have usually some extra features in them, like clock, calender with a reminder function — you don't have to miss any anniversaries anymore — and a pager (using SMS text messaging).

Single-player games are popular in the current 2G and 2.5G handsets. Recently, off-line and on-line electronic gaming have created a quite

Figure 6.1 The Sony Ericsson 3G phone

remarkable software industry with worldwide gaming revenues exceeding that of Hollywood movies. The latest trend in gaming is to connect gamers into networked games. Here 2.5G and 3G allow multiplayer on-line gaming sessions that will bring about further evolution to the gaming industry. For the telecoms industry, networked mobile games will result in more high capacity data traffic.

6.2 The Swiss knife or all-in-one device

Today most terminals with their own telecommunication capabilities have quite similar functionality in them. Some current handsets have more advanced features depending on the preferences of targeted end-users. Some users prefer to have small units to fit it in a purse for example, others require shock-resistance and a dust/water-proof shell for environmental and working condition reasons. An external antenna is an absolute must when the set is operating at the outer limits of the nearest base station. Trying to fit in all features and satisfy all possible needs is of course always a compromise. Needs and preferences in general and how to identify them are discussed in Chapter 3 on Segmentation.

There are already some quite interesting attempts to address the specific needs of some special users. Mobile phones have been introduced with QWERTY-keyboards for those who like to chat with friends with SMS; units with stereophonic radio receivers for music lovers; smart phones with full-scale personal information management software, devices with e-mail and Microsoft Office suite file viewers; camera phones, and gaming phones, as well as Java-supported units that allow software to be installed and the phones to be customized. With the rapid expansion of the abilities and features of phones, soon it will become almost pointless to describe every variation.

Similarly, PDAs have expanded and developed with phone functionality, GPS (Global Positioning System) access, W-LAN connectivity, etc. With some approaching the converged device from the phone industry side, the functionality is on the phone side and units tend to be operable on one hand. Others approach the converged device from the PDA side which typically have larger screens but tend to be operated with two hands.

It is clear that 'one size fits all' does not apply in the future any more than it does now. On the contrary, when the wider selection of services is available the manufacturers will have to design specific models for specific end-users. However, all wireless terminals will most probably have some

Figure 6.2 Digital convergence of portable devices

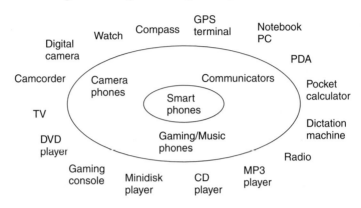

common and basic functions — as they seem to have already — thus the comparison to the Swiss army knife is valid.

6.3 Custom-use devices

In the future, we will have numerous devices with wireless connectivity just like our daily gadgets today all tend to have built-in microprocessors. As our various handheld devices and gadgets become connected, we will also learn to consider the devices by their cost and utility, not by connectivity. In some cases, people will carry multi-purpose pocketable devices that do almost everything reasonably well. Much more often, however, we will be likely to own several devices that are optimized. We may have a gaming device which connects to the 3G network but is not a phone. This device would yield a lot of utility from a powerful gaming interface but we would not carry it with us everywhere like we would the mobile phone.

We may have an Internet surfing device that may have voice ability but is not primarily a phone, and we would use it for portable Internet surfing. Separately, we could have our work datacoms instrument, the logical evolution of the PDA, laptop computer and cellular phone. This would be moderately expensive and include the ability to handle our office data applications and provide secure access to our work data. Beyond all of these we would, of course, have our basic mobile phone and messaging device in a small convenient package. That mobile phone would be selected by its additional features depending on personal preference. We expect most mainstream phones to incorporate the simple snapshot camera utility within a few years. For those with a more serious interest in photography, more technically advanced camera phones will also emerge that will rival mid-range digital cameras.

The future mobile device will be more advanced than current phones, and its evolution will be similar to that of laptop computers. With each generation, an increasing amount of high-cost options and add-on features will become built-in standard features. Thus we can expect that the 'standard phone' in the future will have plenty of PDA functionalities and various models will allow us to include other features we want, such as built-in cameras, music players, game interfaces, etc. We will still have much more advanced game devices than the gaming ability of our basic phone; significantly more advanced digital cameras than the snapshot camera built into our phone, and much more sophisticated music systems than those built into our phone.

6.3.1 The PDA

The PDA or personal digital assistant is becoming a common business tool, especially among all who work closely with technology. A PDA typically has a relatively powerful microprocessor, built-in memory and memory storage ability to handle office software such as word processing, spreadsheets and e-mail applications. PDAs usually have calendar and To Do list functionality and most PDAs also have stylus-based data entry — a keyboard is not necessary.

The PDA is a very useful tool for those who use office software on their computers and want portability and remote access to the applications and their data. PDAs can replace laptop computers or act as more portable versions that can easily be used in trains, airplanes, taxis, etc. Most current PDAs are stand-alone units with connectivity to a PC via cable or infrared, and latest versions are adopting Bluetooth as built-in communication facility as well as W-LAN and/or cellular telephony connectivity, often via expansion card or other accessory.

The PDA is a natural step in the evolution of personal computers growing ever smaller. The trend started with the first PCs that had large desktop cabinets and bulky monitors in the early 1980s. Towards the end of the 1980s, smaller PC cases appeared and the first portable PCs were introduced by Compaq. These were still the size of small suitcases. By 1990 the portable PC had given way to the laptop — a briefcase-sized PC which tended to

Figure 6.3 The Nokia communicator

have monochrome screens — and then by the mid 1990s the notebook PCs with colour screens. Towards the end of the 1990s, the desktop PC had gained flat screen displays making the desktop PC also again smaller in physical size. All along, while the PC grew smaller, the processing power, storage ability and memory capacity of PCs grew at a phenomenal pace. The laptop/notebook versions of PCs of any given generation tended to lag behind the performance of desktop units, yet for most office application uses several higher-end portable computers could equal a typical desktop of the same generation.

From the early 1990s the concept of the palmtop computer or handheld computer became viable. Building upon the earlier concepts of PIMs (personal information managers), which were mostly electronic schedulers, PDAs added true computing ability and software suites compatible with most mainstream desktop computer software. While the processing power, screen size, memory size and storage ability lagged behind those of laptops, the utility of 'pocketable' portability and considerably lower cost than that of notebook computers soon built a market for PDAs.

In 2002 the worldwide population of PDAs was estimated at about 30–35 million, growing at about 10–12 million new units annually. While these numbers are quite significant in contrast with the approximately 125 million existing notebook computers or the 600 million existing personal computers overall, the PDA population must be considered a niche market when contrasted with the 1.1 billion (1 100 million) cellular phones around the world, with new handsets sold at over 400 million annually.

When considering the 3G services and applications opportunity, the PDA can be seen as an interesting opportunity, but it is likely that better chances exist for porting PC-based software and applications users to the PDA, than trying to migrate cellular phone-based applications, solutions and users to the PDA.

To put it in another way, the PDA is a very attractive productivity tool for, mostly, white-collar workers to extend their desktop or notebook computer work and functionality to be readily portable. Such workers form a significant part of the mainly office-bound work force. For these users the connectivity provided by 3G (or 2.5G) will provide considerable utility by giving continuous access and connectivity and thus expanding the benefits of the PDA.

The rest of the population — a much larger part of it — is likely to purchase specialist portable devices of less cost and more targeted functionality, such as game devices, music devices, snapshot cameras, gadgets for the car, etc. These too will soon have 3G connectivity but typically the average housewife has little need for office utility software such as Powerpoint and Excel.

6.3.2 Digital camera

The traditional film-based camera has been a major gadget for several generations. Digital cameras appeared and during the 1990s their price/performance became competitive with film-based cameras. During 2003 it is expected that the digital camera sales will grow past that of film-based cameras. As digital cameras do not have to handle film they can be remarkably small, and already phones exist that are the size of a credit card. The functionality of the snapshot camera can easily be build into a mobile phone as numerous such phones already exist. The evolution from SMS text messaging to MMS multimedia messaging is expected in the industry to ride very strongly on phones with built-in cameras.

It should be noted that the film-based camera is likely to exist in professional circles for a considerable amount of time. The cellular phone plus digital camera combination is likely to be a compromise on digital camera features, bringing only very limited camera ability; hence separate digital cameras of more advanced features are likely to sell in similar numbers as do such cameras today into the foreseeable future. The cellular phone with

Figure 6.4 Video phones will become more and more popular with UMTS offering a range of services. The miniaturisation of camera technology is driving the market towards digital imagery that is part and parcel of the mobile phone package. In the not too distant future all UMTS phones can be expected to include a digital camera for either still or moving pictures. This example is of an existing NTT DoCoMo 3G FOMA video phone from NEC.

built-in camera is likely to replace throw-away cameras and very inexpensive snapshot cameras. There are likely to be few zoom lenses, no powerful flash equipment, and very limited ability to attach any camera gadgets. The devices will be primarily phones with a secondary camera ability. Most serious amateur photographers will prefer to use more advanced digital cameras.

6.3.3 Gaming devices

Another type of 3G device with a specialist use is the gaming device. As a continuation of the Nintendo, Gameboy, Tamagotchi and similar handheld gaming devices, and building on the vast gaming generations that grew up with Sony Playstations, X-Boxes and the PC based games, the 3G phone provides a powerful platform for next-generation gaming. The screen size is relatively small, but mostly colour, giving similar or better screens than the handheld gaming devices. The keypad functionality of a mobile phone can easily surpass the amount of keys and functionality that can be built using the Playstation or similar keypads. The significant added benefit is the connectivity, allowing game play against other humans, rather than against the simple algorithms of the game itself.

The 3G offering at Manx Telecom on the Isle of Man featured the famous PC based network game of Quake. Quake is a three-dimensional virtual game where one shoots monsters and other opponents. Various versions of Quake exist that allow play alone or on PC networks against other human opponents. While many may doubt how a virtual reality game might play on a small screen 3G device, Quake actually worked very well. The gaming industry, already worth more worldwide than Hollywood movies, is very eager to expand into 2.5G and 3G gaming. The early mobile gaming industry itself is estimated to be worth $100 million in 2002 and growing very fast to form a significant part of the $30 billion dollar global gaming industry.

When thinking of the target audience for gaming devices, the most likely targets are late teens and young adults who have already had considerable exposure to the previous generations of gaming devices, platforms and games. The gaming hardware today costs in the $250–$400 range, and individual games cost easily $20–$40 each. When placed into that context, a 3G phone with enhanced gaming ability in a $300–$400 cost range is by no means an impossible cost, in particular if the games can be offered more cheaply than those that are console based.

Where the game owners do not have to press CDs and DVDs with advanced copy protection schemes, package them into boxes and ship to

Figure 6.5 Nokia N-Gage gaming phone

game stores, there are remarkable savings that can be achieved through electronic delivery of the game directly to the handset. Through the advent of pay-per-game and sponsored games, etc., the 3G phone is likely to become a very compelling device for consuming games in the future. Furthermore games are played increasingly by older users. In the USA the average age is a surprisingly high 28 years. Because of the huge installed base of users, when 2.5G and 3G phones form the majority of cellular phones perhaps in 2004–5, the game developers are likely to evolve their best games to function first on the cellular phone, and then only if the popularity warrants it, also release CD based versions for game consoles and the PC.

6.3.4 The credit card

Perhaps a surprising item to consider in this context, but worldwide there are about 2 billion credit cards, of which the most popular is Visa with about 1.3 billion cards. Most credit card users have more than one card, still there are many credit card holders who do not have speficially a Visa card. Most commonly these would have a Mastercard or American Express card. Worldwide the total credit card user population is about 1.5 billion users. While credit card use and the number of users keeps growing, mobile phone user numbers grow much faster. As the reader already knows, there are about 1.2 billion mobile phone users worldwide. Very soon, probably within a year, there will be more mobile phone users than credit card users.

As it is possible to process payments on mobile phones, this population becomes a very significant threat and opportunity for the credit card industry. Numerous partnerships and joint projects have emerged around the world, with the most famous being the various Korean mobile phone subscriptions which also are branded credit cards. Your mobile phone is also a Visa card.

6.3.5 GPS devices

GPS or the Global Positioning System was originally a military system for accurate positioning of military assets developed by the US government. The technology was adopted for civilian use and now many handheld devices exist that give positioning information, maps, etc. Some converged devices have emerged combining a PDA or smart phone with GPS.

6.3.6 3G modems

A special class of terminal for 3G is the cellular modem. 3G modems will be targeted almost exclusively for (laptop) computer use. Initially these will be single mode, i.e. work only on 3G networks, and they will be additional accessories for laptops. Soon multi-mode modems will emerge combining 3G and 2.5G/2G cellular connections as well as W-LAN (wireless local area network) i.e. 802.11 connections. Eventually it is likely that some high-end laptop computers will incorporate the modems directly into the laptop design. The design of a 3G modem is considerably simpler than that of a 3G phone, as the modem will need only the data side of 3G, with traditional voice/phone capabilities.

6.3.7 Custom devices

There will also be various special-use terminals that serve niche markets and their particular needs. For example, the big parcel delivery companies use various parcel information logging and tracking systems that tend to be wireless. These are mostly custom-built handheld devices with, for example, bar-code readers and signature pads, etc. These will increasingly be built to take advantage of 3G. During early 2003, Nokia announced a remote digital camera with a built-in mobile data ability. Such cameras could be installed for various monitoring uses and accessed through the mobile telecoms network. Other such specialist devices can emerge in any industries such as the medical industry, security services, etc.

6.4 Automobiles

Perhaps the most unusual 3G device or terminal is the automobile. While extremely 'mobile' as a phone, the car obviously is not by any means pocketable. We need to keep in mind that the car of today already consists of countless digital processors. There is much added utility to be had from connecting that capacity with central units and a network, from safety, security and navigational needs onto the entertainment of the driver and separately that of the passengers.

The development in automobile telematics is not driven by the telecoms industry, but rather by the big automobile manufacturers, who have seen the profits in their industry dwindle over the decades. Each pursuing the target of manufacturing the world's most intelligent car, all major manufacturers are speeding up the development of data processing and telecommunication ability of the car. The players are all the biggest manufacturers from Daimler-Chrysler, General Motors and Ford to Volkswagen, Fiat, Toyota, etc. Each of the luxury brands such as Jaguar, BMW, Mercedes Benz, Cadillac and Lincoln want to be known as the smartest car. Mercedes Benz came out with their first 3G car in 2002 which was used to trial 3G services on the 3G network in Monaco.

What can we expect? Navigation and safety features may come to mind for the driver, and service maintenance issues to the manufacturer. There are already many solutions today with existing technologies being deployed to fleets of cars which include mapping sofware, intelligent driving instructions, and remote maintenance of problems, for example dispatching emergency assistance if the car airbag is deployed such as the On-Star system in the USA. These are likely to develop rapidly and take advantage of 3G as the network deployment and network coverage allows.

6.4.1 Servicing and maintaining the car

Probably the most mundane needs seen by car drivers are the service and maintenance aspects of telematics. Car systems today have a lot of self-diagnostic abilities, as there are already several microprocessors and sensors throughout the electronics of the car. It is a small step from these to adding communication ability and then having the car contact the authorized dealer and ask for a time for a check-up or oil change, etc. The system can be improved by integration into centralized calendaring and scheduling systems of the car owner, and so the system could look for a suitable moment in time,

Figure 6.6 The BMW 7 series car

not only good for the garage, but also for the car-owner's schedule. The telematics traffic load to the network would be very slight in such systems, where the car would transmit a few bits of data to the centralized scheduling location. These types of telematics data-traffic arrangements would typically be covered by the fleet of automobiles and negotiated as bulk contracts.

6.4.2 Navigation

The next obvious application is finding our way while driving, i.e. navigation. Many CD-ROM-based mapping systems already exist, and some cars already incorporate GPS systems to bring positional accuracy to the mapping software. From here again, it is a small step to a 3G based telematics system to bring more utility and assistance. The main problem with CD-ROM-based mapping systems is that they do not automatically update based on road conditions, repair works, traffic congestion, etc. A dynamic system that contains real-time information on the road conditions and knows the exact location of the car and its destination, will be able to assist even more than a mere mapping system.

6.4.3 Car security and anti-theft

Car security is another area that can benefit greatly from 3G network services. In some countries where car theft is a significant problem, this could be a very high-value extra benefit and will soon be deployed to most new car fleets, and quite possibly retro-fitted to older cars. The car security can have features enabling the owner's phone to be contacted under certain situations, such as the car moving without the owner (or his phone actually) being present, and the owner given the chance remotely to set the car to stop. This would need to be intelligent, as it could not simply apply the breaks while the car is in motion otherwise it could create a serious traffic hazard and another car could crash into the supposedly stolen car.

6.4.4 Multitasking and the car

As intelligent devices proliferate and connectivity spreads, we will increasingly want to have access to our other digital services from the car. Telematics can integrate the car further and help the driver be more efficient through multitasking. For example, the 3G system could connect to the e-mail of the driver, and read out loud e-mail messages to the driver while he drives to work. With simple voice activation, the service could place read e-mails into categories which need to be replied to, forwarded, or deleted. Just as, currently, many drivers use the driving time to listen to voice-mail messages, very soon people travelling in cars can also unload some of the perennial e-mail overload.

6.4.5 Games in the car

Another application for cars is gaming. Not for drivers obviously, but mostly for children sitting in the back seats. The possibility to play Sony Playstation and Nintendo-type games in the back seat, on screens situated at the back of the front seats, is perhaps the answer to the continuous childrens' questions of 'Are we there yet?' The games could be played independently, or against the other child in the car, or not far in the future, also against players elsewhere in the 3G network, for example a family friend who is on another highway also going to their vacation. Before that, however, cars will probably connect to game servers at hot spot areas such as service stations while the car itself is being refuelled. A new game level might be downloaded at the same time for the next several hundred kilometers or miles of travel.

6.5 More devices that seem like science fiction

The 3G opportunity does not end with pocketable devices and the automobile. Arguably most devices that have built-in electronics and a microprocessor can gain added utility from connectivity. The thought of our refrigerator having a built-in camera that we can call up with our 3G phone to see if there is milk in the fridge, when we are at the supermarket, is one that is often mentioned. Some such uses already exist, such as latest high-end copiers being shipped with cellular phone and data connections to allow the self-diagnostics to call up the copier service centre to report trouble and schedule a maintenance call.

One could argue that the connection for machines can be achieved by fixed lines, but it is very expensive to pull added fixed cabling in office buildings, so if the copier is placed in a room where there is no phone jack or that is already being used for example by the fax machine, a fixed connection is unlikely to be used. However, the cellular phone/data capacity can be built in, and the subscription handled by the copier company, producing a built-in utility to the machine, without any further activity by the user.

More advanced visions of what cellular connectivity can produce have been proposed by various futurists and visionaries. Intelligent clothes are one area, where the electronics of the phone can be built into a jacket or pair of

Figure 6.7 Nokia remote security camera

running shoes, and bendable screens set into the sleeves, etc. Another vision is that of the wristwatch phone, which has been tried many times, but with the current level of miniaturization, that vision is likely to become reality. The ability to combine connections via Bluetooth will also produce new distributed computing-based systems, where the loudspeaker earpiece might be built into sunglasses and the microphone into a button placed on the collar of the shirt, while most of the device would be connected via Bluetooth and sit in the pocket, briefcase or purse. A projection display could work with the eyes, and a projection keypad could make use of any semi-flat surface. There is even talk of intelligent shoes which monitor the state of the feet and adjust elements in the soles to adjust the shoes for perfect fit.

Again, advanced 3G gadgets will not be limited to those serving humans. Locator devices so that we can keep track of our pets are one possibility. Intelligent roads are another, where roads become aware of the traffic load and then communicate with the cars, set priorities for emergency vehicles, etc. The street lighting services are now being upgraded with sensors to detect when the light bulb has burned out, and with a mobile chip, the light post sensor will report burnt out light bulbs. The savings in manual labour with such services can be considerable as early studies in the UK have illustrated. Again, simple versions of all of the systems mentioned here are already in trials around the world.

6.6 Handset subsidies

One of the very significant *marketing* dimensions in mobile telecoms is the matter of handset subsidies. We will discuss the effects of subsidies more thoroughly in the churn chapter (Chapter 17), but the matter should briefly be discussed here as well.

The first issue that all mobile telecoms marketing people should know, is that subsidies can be radically different in any given market. In most markets mobile phone handsets are subsidized. In extreme cases phones are given for free or for a token fee such as for the cost of 1 dollar, but there are countries where there are no subsidies whatsover. These markets illustrate a very healthy mobile phone environment — and no harm to penetration levels — but the behaviour of consumers is more rational than those where subsidies are extreme.

For those who work in markets with subsidies, it may come as a shock that of the world's leading markets that have been at the very forefront of mobile phone penetration for the past ten years, in Italy and Finland — both with subscription penetrations well beyond 90% and still rising — there are no subsidies whatsoever. All consumers pay full retail prices for their mobile phones. This has in no way deterred them from buying ever more expensive and feature-laden handsets. Recently, Korea joined the subsidy-free range, and we expect many more countries will follow during this decade.

At this point we will simply summarize that in the opinion of the authors, handset subsidies are inherently bad in any mobile telecoms market at all times. Subsidies are wholly unnecessary for a vibrant mobile telecoms market as proven by Italy and Finland. Subsidies trigger behaviour where the users do not follow reason. Users will change operators just to get a new handset. They will demand new handsets from existing operators to remain on the network. The users will ignore commitments to long-term contracts. The list of harmful effects is long and discussed more in Chapter 17.

In sum, the subsidy situation presents a classic 'prisoner's dilemma', where often the industry understands that the subsidies are bad, but any single operator cannot find it in its interest to stop the practise unilaterally. Here the operators can lobby the regulator or try to find industry consensus to steer away from subsidies. We expect that in the long run the industry will be able to diminish and eventually extinguish subsidies in most markets. It will not be an easy road, but in such an uncertain world as 3G, perhaps this is one uncertainty factor that the industry should resolve to its benefit.

6.6.1 Device needs

The 3G terminal will have a lot of potential and can be any combination of the above mentioned features and more. The devices will evolve and improve with time. The devices will also have some very unique requirements that we need to be aware of:

A — Always
B — Best
C — Connected

6.6.2 Connectivity

The 3G devices will increasingly need to integrate and connect with other networks and technologies in addition to 3G. In most markets it is expected that most 3G handsets will need to be able to operate also in 2G/2.5G networks. The needs also extend to connect with IR, Bluetooth and W-LAN hot spots and short-range connectivity. Many other connections may be desired, such as access to GPS or digital TV signals. For these purposes Eva Gustafsson and Annika Jonsson of Ericsson Research have proposed the concept of ABC — Always Best Connected, by which 3G terminals always seek the best available connectedness, both from a technology price/performance point of view, and the user's individual preferences. For example while a W-LAN hot spot might be available, a company-owned 3G phone might not be allowed to access it for fear of loss of security, and the company would prefer to pay 3G-connectivity charges to get the full benefits of the security levels provided by data access via the 3G network.

6.6.3 Synchronization

Another rapidly growing need is to provide seamless, automatic and unobtrusive synchronization of various digital gadgets. The calendar on the mobile phone should synchronize with that of the company calendaring software. The snapshot pictures taken with the camera-phone should be moved to the family photo library. The new music bought on the fixed Internet should be easily brought onto the portable music player/mobile phone, etc. As services and applications get more complex, the synchronization needs can also grow. For example a multi-user network game with the ability to play from a broadband Internet connection, and from a 3G phone, should allow the 3G phone user to switch over to the broadband internet connection easily when at home. The game needs to recognize different devices, such as perhaps the game also being played from the backseat of the family car.

6.4 Handing off on handsets

Most users of 3G services will be thinking of 3G as their terminal(s) and associating all mobile service benefits in relation to their handsets and other devices. The 3G terminal therefore plays a critical role in the adoption and attitudes to modern mobile services. The form factor for emerging terminals may be radically different to what we are accustomed. The combinations of Internet speed and the innovation of global IT, telecoms and personal electronics industries will create fantastic new devices for us.

While most of the developments are likely to be positive, perhaps we can learn from Robert X. Cringely's thought: 'If the automobile had followed the same development cycle as the computer, a Rolls-Royce would today cost $100, get a million miles per gallon, and explode once a year, killing everyone inside'.

We don't care. We don't have to. We're the phone company.
— Lily Tomlin

7

Distribution
Marketing and Selling Through Channels

The distribution channel in mobile telecoms needs to look at two separate purchase decisions. Initially a customer buys a terminal and signs up to a subscription. Then the customer will hopefully also use (buy) services available on the network and consumed on the terminal. The issue of handset sales is rather easy to grasp: we've all been to mobile phone stores with rows and rows of new mobile phones. The subscription is almost always bundled with the handset sale, and a SIM (subscriber identity module) card is inserted into the phone at purchase. As both purchases involve a tangible

---

3G Marketing: Communities and Strategic Partnerships Tomi T. Ahonen, Timo Kasper and Sara Melkko
© 2004 John Wiley & Sons, Ltd   ISBN: 0-470-85100-7

item — a phone and a SIM card, the warehousing, shipping, store display, sale and support of these items is quite straightforward.

A more conceptual matter is the sale and distribution of the services in 3G as well as those already in 2.5G. How do you sell a mobile commerce purchase activity? Or how do you sell the ability to view news updates? Moreover, what are the distribution channel issues relating to the sale and delivery of such intangible services? This chapter will examine all of these issues.

7.1 Sales channels

Before we can examine the different issues relating to the distribution channel, let's start by examining what kind of distribution outlets exist or are emerging for mobile telecoms. In the times of telecoms monopolies, the telecoms operators — or what were usually post, telecoms and telegraph (PTT) authorities of the government — started to set up service offices or bureaus where customers could come to pay bills, pick up fixed-line telephone sets, fill out forms such as making an address change, etc. As more customer-service oriented philosophies started to emerge in the 1980s, these bureaus and offices became 'stores' and 'shops'. However, even in some countries today, they more resemble government bureaucratic offices than competitive stores. It was often the mobile telecoms operators who brought more modern stores into the telecommunications world during the 1990s.

7.1.1 Operator's own stores

The traditional model for distributing mobile services has been through mobile telecoms stores and shops. Usually the mobile operator has its own branded stores, and its services are also offered through independent distributors, such as electronics stores, department stores, and specialized non-affiliated mobile telecoms stores, such as Carphone Warehouse in the UK. The traditional stores that sell mobile phones and subscriptions have the best trained staff to handle newer more advanced mobile services, yet they all will need a lot of further training for 3G.

The mobile operator's own stores are a significant distribution channel and also an auxiliary tool for other marketing purposes, such as brand identity and market intelligence. The mobile operator's own stores are the only ones that typically carry no competitor products and services. Any person walking into the company store is a prospective customer. The staff and

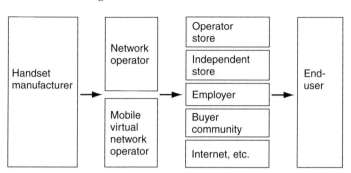

Figure 7.1 Handset sales channels

the store is perceived by the customers usually as the face of the mobile operator — or its parent company if it is part of a larger telecommunications company. This means that a customer can walk in expecting to be able to pay their fixed line phone bill or get assistance with their broadband modem, etc. Customer enquiries at stores can cover a very broad range of topics.

The key to maximizing the benefit of a mobile operator's own stores is to manage the stores and motivate the staff particularly well. The stores serve several functions apart from distributing mobile phone handsets and connecting new subscribers. The store staff will need to be very well trained on the current services that are being promoted and pushed by the company. The store staff will need to be rewarded and compensated for selling services and service bundles and, especially, the staff needs to be motivated to train end-users to be comfortable with the new services. Store staff will need to instill upon the store customers a strong sense of care, so that the customers feel that they can often return to the store, even if only to get information. It is in the store's interest to make sure that its own customers do not walk into neutral stores — which can sell competitor products and services — or even worse, walk into a store of a competitor.

7.1.2 Independent stores

The second type of store is the independent mobile phone store. These specialize in mobile phones and usually sell subscriptions to all network providers and most major MVNOs (mobile virtual network operators) in the market. The staff tend to be well trained and knowledgable. They, of course, know rather well the major differences between the most popular service bundles of the major players, but as the sales staff are sales people, they are strongly motivated by money and various incentives such as sales contests. The mobile operator will need to nurture a good relationship with as many of the independent stores as possible and try to accommodate their needs.

7.1.3 Departments and sales desks of other stores

Many department stores, electronics stores, larger stationary stores, etc., are setting up stands and sales desks for selling mobile phones, subscriptions and supplies. The staff manning these stands tend to cover many areas and will not be strongly focused on mobile phones. There are typically one or two trained mobile phone specialists, but the rest of the staff will be very weak in understanding the finer points of services, differences, etc. These outlets will need simple guides for fast answers. They may be interested in limiting their selection to a few from those on the market and relationship management will be critical in ensuring that any given mobile operator (or handset brand) will be carried in the line-up.

Recently, more varied store channels have appeared such as major grocery chains offering mobile phones, subscriptions and top-up cards for prepaid accounts. In future, it is very likely that a wide multitude of outlets might start to sell individual mobile services. For example, the local newsstand, which sells picture postcards for tourists, may also start to sell mobile multimedia postcards, the local locksmith may sell mobile security solutions, and the car-radio specialist may also offer car mapping solutions as retrofits to older cars. The mobile service distribution is likely to expand very much from that now current.

7.1.4 IT integrators

With business customers ranging from the small to the corporate, a lot of telecoms services are sold via IT (information technology) integrators. These can range from the neighbourhood computer sales and service centre to global IT companies like SAP and IBM. The key is that computer systems tend to involve networking benefits, and increasingly these involve a mobility or wireless element.

Very often, the IT department of a business customer is likely to think of the mobility dimension as a desirable feature to an existing system such as e-mail, or access to the corporate intranet, LAN (local area network), etc.

Figure 7.2 Music industry revenues by channel in 2003

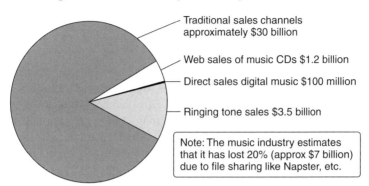

Sources: International Herald Tribune, Pyramid Research and Tomi Ahonen Consulting 2003

These are not thought of as wireless telecoms needs that could be met by the mobile phone, but rather as usability improvements to systems that already exist on fixed computers and laptops, PDAs (personal digital assistants), etc.

Here the mobile operator can find a new sales channel in the IT integrators. If the IT expert tells the business customer that the system works and has been tested with mobile operator A's network solution, the customer is quite likely to want to use that mobile operator for that solution. This is particularly true of emerging technologies and innovative services, rather than that relevant to commonly used technologies such as SMS and WAP today. Also the influence of the IT integrator is greatest when the purchase is a new one, not when the customer already has a terminal and subscription of equivalent ability.

The IT integrator will be needing its own support. These customers will be talking in the computer technician's language rather than the telecoms language. The IT integrator will tend to know the IT requirements of its typical customers but can often be weak on the telecoms and mobility sides of the issue. The support staff on the mobile operator's side, as well as the major technical support documents, should reflect the IT terminology.

7.1.5 The Internet as a sales channel

A relatively new way of processing customer requests, service changes and upgrades is by using the Internet. Not all customers have Internet access, but in Western countries the majority of families and working people do have access through some means. The Internet can be a very cost-effective means to serve customers. As a distribution channel, it is likely to be seen as a threat to traditional shopfronts, especially if any special discounts and offers are provided through the Internet that are not available at the store.

7.1.6 The mobile portal as a sales channel

The newest way of delivering new services, upgrades and features is to use the mobile phone as a delivery channel. This channel is still in its infancy but is likely to grow to be very useful. In many cases, an advanced mobile phone is the ideal interface between individual customer and a service upgrade or installation. The mobile portal and its impact on 3G marketing is discussed in more detail in its own chapter (Chapter 8).

7.1.7 MVNOs

A new phenomenon is the MVNO (mobile virtual network operator) where separately branded phones and subscriptions are sold by companies that do not own a licence for radio spectrum, and do not operate the radio network actually to serve the customers. The connectivity is provided through one of the existing mobile operators. The MVNO may appear in the market place as being similar to the existing network operators, and a good example of a mass market MVNO is Virgin Mobile in the UK. Probably numerous niche market MVNOs will also appear, such as the *Financial Times* MVNO already in the UK. The opportunities could range from familiar mass market brands such as Daimler-Chrysler offering mobile services for its fleet of new cars, MTV offering youth- and music-oriented services, to industry specialist solutions such as medical services offered to doctors and hospitals.

7.2 Managing channel conflicts

When considering the distribution of new mobile services, the first and most obvious area that needs to be addressed by the network operator, portal and/or or service provider, is potential channel conflict. The potential for conflict increases with new traditional channels and also from working with MVNOs. Conflicts can range from pricing and discounts, to servicing customers with complaints, to connections to services. The operator's own branded stores will want to maximize the benefits of their affiliation with the brand. Any independent stores will feel discriminated against if they do not get the same opportunities as the operator-branded stores. The operator's own store staff will feel it unfair if the independents get all the benefits that they do, feeling that there should be some benefit from being the brand's own outlet.

The primary way to deal with channel conflicts is to be fair with the rules and give clear guidance on what the opportunities will be for any level of affiliation with the brand. For example if e-business savings are passed on to customers by the operator if services are bought via the Internet, then similar discounts should be made available also for orders placed through the independent re-seller's website. Upcoming changes should be well communicated to the chain(s) to minimize surprise in the distribution channels.

The shopfront is one of the primary ways that customer experience is created and maintained for the service. Operators should spend considerable time and effort to ensure that this contact is well received by customers.

7.3 Selling new mobile services

It is often difficult for consumers to come to grips with new ways of consuming traditional services. It may be useful to box or 'package' a service offering, even if no actual devices or software, etc., are needed. The box could contain some user information, a user's guide and possible pricing information. This could help in offering any strictly intangible services such as setting up a mobile banking service or insurance or gaining access to mobile games or music. Arguably the service should be completely deliverable via the mobile phone and, for example, a WAP (wireless application protocol) interface, but users may be unwilling to 'surf' looking for instructions for random services, and may be quite willing to trial such services if the instructions come in a box. Over the long run this is likely to change as users become ever more 'm-literate', but early on it may help with launching new services. It is also a good tool to use for content providers who are introducing mobile versions of their services.

7.3.1 Bundling an m-component

For many industries a new mobile service can provide an added utility or dimension to an existing service. For example, a locksmith provides services for locks and security. A mobile service option could include services like setting the air conditioning or heating of the home remotely while still driving the car, setting the home security system on and off remotely, having remote cameras that can be accessed from the mobile phone, etc. In these cases the primary service is still the locking and security service provided by the company installing locks and security systems, while the mobile component is an addition. In such cases, the locksmith would be an unlikely source for selling primary mobile subscriptions and handsets, but could well have a stock of premium phones and the ability to provide the service upgrade that could be done in the store for a customer.

In another example, a pop star could bundle a set of ringing tones together with the early release versions of their latest album on CD as a sales incentive to climb quickly into the charts. The packaging of the new album cover would feature the information on how to get the selected ringing tones, which would naturally be the band's current and recent hits. Again the music CD is the primary sales item, the ringing tones a bundled component. The range of options is almost unlimited. For example a camera-accessory to an MMS phone could be bundled with a give network's deal of, for example, 100 free MMS messages, etc.

7.3.2 Soul of the store sales rep

A significant key to how mobile phones, accessories, subscriptions and services sell is the actual in-store sales representative. Often the sales rep gets burdened with numerous overlapping goals and objectives. It is imperative to understand that the sales representative is driven by money or equivalent benefits. The sales representative in the store will sell that service and handset which gives him the best return. The benefits include sales commissions, sales awards, and hitting bonus targets.

The real daily examples from the in-store sales staff can be demoralizing. For example, if the sales rep knows it takes about 15 minutes to explain GPRS to a customer but only 5 minutes to sell the service, and the sales rep is not compensated for his additional 10 minutes of end-user training, then the sales rep will learn the fast arguments to convince the customer to purchase GPRS and sell three services in 15 minutes — collecting three times the sales commission and/or bonuses — rather than bothering to teach one customer how to use the service. The damaged party is naturally the mobile operator, who gets unhappy customers and then receives plenty of help-desk support calls to the calling centre, resulting probably in much more time spent with the same customer. It would have been much more efficient to teach the customer at the first instance in the store.

The problem is not that the sales rep is irresponsible or malicious. It is simply that the store manager or network operator did not construct a suit-

Figure 7.3 Relative strengths of operators and third parties

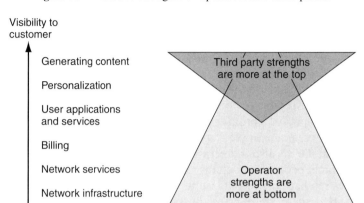

able motivation system to ensure that each GPRS customer is clearly shown how to use the system. Sales reps are very predictable and any sales plans and targets should be run by a small trusted sample of current senior sales reps who know the storefront situation. These sales reps should not be punished for giving their own opinions and they also should not be used to try to re-educate the other sales reps, as that will never happen. A sales rep is driven by his greed and the perceived value of the various benefits that are available. A quick sanity check with the trusted senior sales reps will save a lot of headache and help to ensure that any new promotions, sales plans and targets can reasonably be met. We will discuss the sales representative in more detail in the Chapter 13.

With stores that are owned and controlled by the network operator there is little chance for competive offerings. With independent chains of stores, such as Carphone Warehouse in the UK, the problems of motivating the sales reps are magnified as there will always be competitive offers and promotion schemes. Any new promotions and sales campaigns have to be considered also against what the competition has on offer for that given distribution chain.

7.4 Information flow

The mobile operator will have learned to support their own stores with sales support materials and various forms of training, tools, etc. More than likely it has also discovered that most other channels require the same kind of support to do a credible job. The sales channels are the first line of information for many kinds of information, in particular about competitors. The mobile operator will need to harness the information in the sales channel and motivate the sales staff at all outlets to collect and report relevant information. It is important to remember that in addition to information about sales, a lot can be learned from instances when a sale does not happen. Perhaps the prospective customer asked about some service that is a brand name of a competitor's product. Maybe the prospect was asking for something which does not exist yet — providing priceless information for product development. Or, as also frequently happens that the customer, was asking for something, perhaps in a clumsy way, that would have been available if your sales rep had just recognized it. This is the most important lesson to learn — that there are gaps in the sales force where a willing customer would have bought your product, but the sales rep was unable to identify the need and make the sale.

The distribution channel staff will need to be encouraged to report all relevant information. They will not be naturally inclined to understand how broad this request is unless they are very well trained. The store staff has very many different types of rapidly changing information to master and thinking about what might be important to the mobile operator is far from their minds. For the system to function well will require not only recognizing, but also rewarding the people who bring in the information.

7.5 Warehousing, shipping, inventory

Another common aspect of the distribution channel is the delivery chain involving warehouses, physical shipping of goods and managing inventory. For the most part in mobile telecoms, this is not a major competitive factor. With mobile phone handsets there are delivery, warehousing and inventory issues and with the short lifespans of mobile phones there is a need to ensure that the inventory does not carry obsolete or obsolescent models. The stores and warehouses need to make estimates of likely sales and the mobile operator can assist in these kinds of forecast, etc. Still, mobile phones are rather small items of relatively high value, so their physical distribution issues are less of a critical factor than, for example, in automobile, furniture, etc., types of industries; nor will the handsets represent a perishable good with a critical life span such as flowers or fresh food.

With the network services there are potential delivery problems but these are called congestion on the network. There are dimensioning techniques and tools, and there are various traffic shaping tools as well to help resolve network congestion issues, but these fall into the side of network management and are beyond the scope of this book. We recommend the book by Laiho, Wacker and Novosad entitled *Radio Network Planning and Optimisation for UMTS*.

7.6 Distribution as an end

This chapter looked at the distribution channel as it relates both to the delivery of 3G handsets and the delivery of 3G services. The mobile operator will be learning to manage an ever increasing range of distribution channels from the traditional operator store to both fixed and mobile Internet delivery. The biggest challenge will be to handle the information flow and prevent channel conflicts.

While the distribution of wireless services is not as massive a logistics headache as the delivery of, for example, perishable goods, the distribution in 3G will be an important marketing function. As with most of the evolving aspects of modern marketing with 3G, so too with distribution the key lies in managing relationships. Here we can take advice from Brian Tracy who said said: 'Your social intelligence — your ability to get along with others — is the single most valuable asset you have. Eighty-five per cent of what you accomplish in life will be a direct result of how well you get along with others'.

> *In the Internet business, profitability is for wimps.*
> — Mike Doonesbury (comic strip)

8

Portals
The Window to Mobile Services

The portal is the first experience the user gets of mobile services. The portal is how the user experiences the mobile service offering and the portal is the service provider's way of establishing its brand. Increasingly with 3G, the portal will be seen as equivalent to the actual service offering, and any content, applications and services will be considered synonymous with the portal.

The mobile portal is a very new phenomenon and is not, therefore, well understood by many involved with marketing advanced mobile services. This chapter scrutinizes portals and their obvious crucial role when talking

---

3G Marketing: Communities and Strategic Partnerships Tomi T. Ahonen, Timo Kasper and Sara Melkko
© 2004 John Wiley & Sons, Ltd   ISBN: 0-470-85100-7

about marketing 3G services. First, however, we give you a short introduction to the history and, hence, definition of a portal. Then the importance of content is analysed further within this chapter.

8.1 Defining portals

The term 'portal' is derived from fixed Internet usage. The word itself means doorway or access. Portals are mostly used as the first (default start) website to appear on your screen when starting your internet browser. A portal therefore, is the opening web page of the website of the Internet service provider (ISP). These sites are used as an entrance to the Internet — hence the term portal.

In the early Internet days, when navigation on the Internet was still a challenge, the early mass market Internet access providers evolved from what were called bulletin board systems (BBS) such as Compuserve and AOL. Several other of today's popular Internet portals started as simple search engines, such as Yahoo!, Altavista, Infoseek and Lycos. With the rapid take up of the Internet, these portals were followed by the Internet browser makers' own portals, the most important of which being Microsoft's MSN riding on the success of Microsoft Internet Explorer. Many of the incumbent fixed telecoms operators around the world and their major national ISP competitors have succeeded in launching successful and frequently used portals. Various e-businesses (e.g. Amazon), online newspapers and business sites invaded their own niches with portals as well, and some of the most successful ones also show respectable page impression numbers.

8.2 3G portal categorization

Over the past three years mobile portals have emerged. 3G portals are superior to fixed Internet web portals in many ways. The most crucial aspect is the authentication of the user. As a result, the mobile-portal owner can achieve interactivity and a high security level and hence the possibility to bill and use e-payment.

8.2.1 Different types of mobile portal

There are several different types of portal services. The main types are mobile intranet/extranet portals, customized infotainment portals, multimedia messaging services portals, mobile internet portals, location-based services portals. The UMTS Forum 2001 Report No. 16 defined a set of different 3G portal services that are listed below:

Portal types according to UMTS Forum
- Mobile intranet/extranet portals;
- Customized infotainment portals;
- Multimedia messaging services portals;
- Mobile internet portals;
- Location-based services portals.

8.2.2 Categorization

Most portals have a surprisingly simple concept of how to get users to visit the site time and time again. In addition to the navigation (search engine), you often find news (weather services, sports, business), categorized links (useful link collections like travelling, sell your car, get a new job, etc.), communication (online chat, forums, dating services), entertainment (online games) and online shops (together with partners). The trick is, of course, to try to keep the visitor on your site as long as possible and have him come again and even recommend your site to friends.

8.3 The 3D rule for mobile portals

Your 3G portal must, of course, meet different user demands in comparison to web portals. Mobile surfing — or 'murfing' — is somewhat different to websurfing. Surfing as we understand it within the fixed Internet world is a term associated with the drifting within cyberspace in near-effortless search of information, entertainment or something new or different. Typical of surfing, we have time to land at different websites, explore and follow through with links. When surfing we often follow interesting links out of curiosity. In the course of fixed Internet surfing you might end up somewhere other than where you were intending to go in the first place.

With the mobile Internet there is no tolerance of endless and random surfing. Murfing is very expensive and in most cases the mobile phone user is in a hurry when starting to murf, where the user clearly knows that they want to reach a certain point or piece of information in the shortest possible time.

8.3.1 What is murfing

There is a 3D rule to murfing. For a portal to provide value to the user, it needs to address three Ds:

Duration (time)
Display (terminal)
Demand (truth)

Duration (time)   Time has two dimensions in 3G. First of all a user's mobility often demands that the required information be produced quickly. A 3G-portal visitor does not show the same patience as is the accepted norm with traditional fixed Internet webportals. Early studies have shown that the murfer grows intolerant at delays of 2 seconds and may stop the activity, whereas on fixed Internet surfing, the occasional delay of dozens of seconds is happily accepted. Therefore navigation and site construction play an important role. Secondly, the actual time of day might play a role in what a portal will offer you.

Display (terminal)   Display can also be described with 'dimensions'. Some of the most restrictive factors for a portal's structure are the constraints of the terminal that the given user happens to be using. On the fixed Internet the web pages tend to be built to display standards such as SVGA (Super Video Graphics Array), while mobile phone displays vary widely even among phones from a given manufacturer. The limited display size has to be always kept in mind when designing a portal.

Demand (truth)   The accuracy of the demanded information is maybe the most important factor in creating a successful 3G portal or 3G-portal dialogue. There are enormous challenges in almost 'guessing' what kind of information the user is murfing for, presenting the possible results in a concise way and omitting any superfluous or unwanted data. Search engines will be expected to work even more accurately according the principle WYDIWYG or 'what you demand is what you get'. The user is not expected to navigate through dozens of different sites of varying degrees of reliability and relevance before reaching the preferred destination or information.

It is easy to understand that in the first stage of mobile content development, the portal design should revolve around the question 'what information is critical when the customer is mobile?', such as information and infotainment, search services (e.g. nearest Chinese restaurant) as well as communicative services and entertainment. With entertainment, however, pricing becomes increasingly important. Here again, operators should orientate on existing highly successful services such as SMS. The basic law of supply and demand meeting at the optimum implies a correct price level.

8.4 Personalization

A correct identification of the user has been one of the Internet's biggest challenges. So far, we have had to rely on usernames and passwords to provide access to registered services. Even though Internet protocol allows for identifying individual devices, the identification of a user by a device's IP address is not a very reliable method. There are issues such 'fake' IP addresses, dynamic IP addresses, and instances where the IP address stands for the device, but not necessarily the human being behind it. One partial solution has been the use of 'cookies' installed on computers to store data on the user and to identify the user for the service. Cookies suffer from the same problem — where computers are shared, so too will a cookie be easily confused about who it is who wants to access a service.

When registered on the fixed Internet today, we are often given the chance to design a portal according to our wishes and choices. For example, Amazon.com remembers and coordinates requests and orders and automatically recommends other similar items. We've all seen the type of automated recommendations, such as 'people that bought this book also bought books by the following authors'. These concepts will become multi-dimensional in 3G and personalization becomes easier.

8.4.1 Authentication ('intelligent' portal)

The mobile phone SIM (subscriber identity module) card has evolved and gained functionality and memory. With 3G, the SIM card becomes highly personal and could be seen as a quasi-thumbnail of the user. Partly because of the SIM card, and also because we don't share our mobile phones as readily as we share computers, the determination of the user's identity by the device (i.e. its 'memory'), along with passwords like the security PIN (personal identification number), becomes highly accurate.

Portals accessed via a certain device can already display the explicitly expressed or previously frequented areas of interest. Portals adapt and learn more about the user every time the user requests information and the portal can respond to requests with increasingly accuracy. Therefore 3G portals could be described as 'intelligent' since they permit high interactivity with a particular individual.

A separate but very significant benefit is that the authentication provides a basis for secure billing. As the mobile Internet is naturally able to handle payments, this provides a key for migrating content from the fixed Internet.

8.4.2 Timing ('instant' portal)

Timing has another use. The content and structure of a portal might vary according to the point of time. According to our preferences, a portal might offer us access to professional content during working hours on weekdays, and then offer our preferred selections of entertainment in the evenings and at weekends.

Some information has value only when obtained urgently or in near real-time. Typically such information would be a sudden change in the value of shares in a stock portfolio. As with the mobile Internet the users are constantly connected and hardly ever switch their device off, they can be reached when pertinent information or news happens. This type of service, based on critical timing, makes for an interestingly unique selling proposition.

8.4.3 Positioning (portal 'to go')

The mobile phone is a highly personal device and given the same care as are keys and wallets. We take our mobile phone wherever we go. The mobile phone's location can be detected by the network. This detection of a user's whereabouts means that — to a certain extent — the user can expect the portal to adapt and respond to where they are. An intelligent portal could detect, for example, whenever we are at the vicinity of our office, and automatically offer us the work-related links and set-up at that location, even if we happen to come to work on a weekend or stop by the office briefly on a late evening, etc.

8.4.4 Pull versus push (portal 'on demand')

A last way to personalize the portal is have the user determine if and what information is sent automatically when an update becomes available. The traditional way of considering fixed Internet content and early mobile Internet content is the 'pull' service. With pull services we seek information, much like looking in the Yellow Pages or reading a newspaper. We have to actively go and seek the information or service. When content and services are repeatedly accessed, or their access can be predicted, a more valuable service type can be created, called the 'push' service. With push services, the user gets the information offered when it happens. This is similar to listening to the radio and getting a news flash. With the 'always on' character of 3G, a wide range of push services can be created, delivering much added value from the mobile portal.

8.5 Open content policy – a decisive battle over 3G's success

One of the decisive success factors of the fixed Internet is its open policy, or in short, free global access to 'all' information. Of course with the advent of intranets and extranets, not all content on the Internet today can be accessed by everyone. The philosophy, however, of free access to everything still prevails. WAP on the other hand, tried to give exclusive content to its users from a strictly controlled and limited selection of service and content providers. This has been one major reason for its failure. In the very early days of the mass-market Internet there were similar 'walled garden' Internet service providers. By the mid-1990s all such service providers had to open access to the full Internet or perish.

There has been some suggestion by some 3G operators of attempting to recoup 3G investments by offering only their own content. This we feel is very short-sighted, will result in dissatisfied users and may result in a backlash once open access is made possible by some player in the market. The mobile operators should learn from lessons in Korea and Japan, where all operators offer their own branded portals with 'official site' content and services, and through the mobile operator's portal also access to the rest of the universe of mobile content providers as 'unofficial sites'. The operators tend to offer their extra-value service billing features only to official sites.

8.5.1 The more services, the more money for everybody

In some countries, network operators still pursue the philosophy that it is their network and their customers and that they want to be the sole service or content provider to their clientele. Operators can also be very restrictive to other businesses that try to access their customers. What these operators do not realize is that the more interesting services you can offer, the more revenue there is at stake for all parties. Many services, for example the logo and ringing tone SMS services, have been word-to-mouth products that became successful only when a large variety of service providers entered the market.

There is a temptation to use the exclusive access as a means to force a customer to select from a given portfolio of services. We very strongly advise against such strategies, and recommend that the policy should be open. The argument for open access was proven by the early success in Japan and Korea versus the European disappointments of early WAP. Furthermore, portals will no longer solely reflect a single company's views and contents. Content partnerships will increase and portals will increasingly focus on personalized subject matter and address a single users's wants and needs. This leads to increasing cooperation and transfer along the value chain. In addition to mainstream portals, there will also be marketplace for specialized portals and content providers. The operator's unique sales proposition is not in the creation of content and they should encourage and acclaim this division of labour and team play. The more compelling the content the more traffic and money to share.

8.5.2 Price strategies: skimming versus penetration

There are different approaches to pricing, namely skimming and penetration strategy. Price skimming (or premium pricing) is a product pricing strategy, where a company tries to take advantage of (a usually small group) of innovators that are willing to pay the highest possible price for a product or service. Usually, the price is then lowered to attract other customer groups such as early adopters when technologies mature, acceptance grows and competitors emerge. The need to lower prices is important as imitators tend to be much more price sensitive than innovators.

The usefulness of a price skimming strategy can be tested with the following questions:

Questions about price
- Do customers associate high price with high quality?
- Are innovators willing to pay higher prices?
- Is the number of innovators critical?
- Do high prices lead to more fierce competition?
- How great are the barriers to market entry?
- What is the price elasticity, i.e. is there correlation between lower price — more customers — economies of scale (costs per 'unit')

Due to their strong customer base and customer loyalty, operators might be seduced by the thought of 'monopolizing' the selection of partners. In the early stages, the enormous infrastructure costs and internal pressure of drawing profit seems to validate this type of approach. We think that this *modus operandum* is not only short sighted but could even be a death blow for the development of the whole business. Consumers expect similar openness and easy access to a wide range of differenet services from the mobile Internet as they get on the fixed Internet. Consumers clearly will not welcome another 'walled garden'.

Price penetration, on the other hand, means launching products at a low price level in order to gain a quick market share, or in 3G, a critical mass of users in a relatively tight time schedule. The penetration strategy goes along with extensive marketing measures and is very cost intensive. Even thought operators might be appalled by the thought of entering markets with a low profit margins, this is the only way to ignite the whole 3G business. We believe that the business needs this spark and investment. We will discuss the pricing strategies in more detail in the Chapter 14 on Tariffing.

8.6 Revenues and advertising

Apart from generating costs, portals can also produce revenues. There are direct means and steps to generate revenues from portals and there are also ways to increase customer loyalty or customer intelligence and harvest revenues later.

8.7 Collect customer data (registration)

Many companies offer free services in exchange for registration, for example free e-mail accounts or club memberships often including some free gifts as well. As we saw from the previous chapters, it is vital to understand your customer well for your segmentation to succeed. For many ISPs, their own portal is by far the most important tool to connect with the customer and to get to know the customer better. With the fixed Internet there is the challenge of attracting truthful statements from visitors who register. As registration is annoying to the customer even on the fixed Internet, it will be much more so on the mobile Internet and user data should be collected unobtrusively.

8.7.1 Advertising

The classic means of fixed Internet advertising is of course banner advertising. Banner sales have not developed into a major source of revenue for the fixed Internet portals. They could, of course, mean direct money, but are distracting to the user and are increasingly ignored. In a 3G setting classic banners as such will be even more unwanted due to the display size and the time sensitivity. In terms of how annoying they are, banners are only surpassed by pop-up ads, which even on the fixed Internet can totally ruin the surfing experience. While users are willing to put up with a considerable amount of advertising on the fixed Internet, on the mobile internet the users will be much more critical of ads, first and foremost because every minute of air time is billable time.

The mobile portal will need to select its advertising methods carefully. Choosing the wrong form of advertisement in order to gather short-term revenue can easily result in severe backlash in the 3G customers' behaviour, even to the degree of seeking to change operators. The guiding principle must be that the users 'opt in' or actively accept receiving ads. The mobile portal can offer remarkable targeting, location-based triggering, and time-sensitive dimensions to ads that other adverstising media simply cannot provide. The introduction of ads will need to be gradual and the mobile operator and the mobile portal must remain alert to any developing customer backlash.

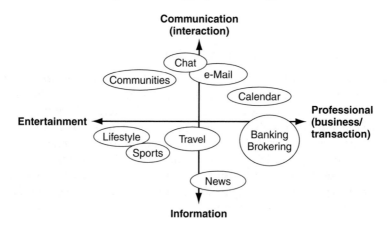

Figure 8.1 Mapping of service types

Table 8.1 Comparison of fixed and mobile Internet in 2004

Aspect	Fixed Internet	Mobile Internet
Devices in use	Fewer PCs and PDAs	More browser phones
Cost of device	More	Less
Sell annually	Less	More
Messaging users	Fewer e-mail users	More SMS users
Cost	Less	More
Preference	Preferred by older	Preferred by younger
Content	Mostly free	Always costs
User willing to pay	Reluctant	Assume everything costs
Billing	Cumbersome	Built-in
Micropayments	Very difficult	Easy
Paid content	Is big problem	Is no problem

As a separate form of advertising revenue, we believe that the sponsorship opportunities with mobile services are considerable, and the mobile portal should develop these methods. Sponsored services can include direct sponsoring such as McDonald's, sponsoring the tourst maps on 3G phones for free, with each map of course showing the locations of the nearest McDonald's hamburger restaurants as well. Similarly a beer company like Heineken might sponsor discount tickets to a bar or night club, and Nike might sponsor the free trial of the newest skateboarding game. The opportunities are endless. We think that eventually customers' acceptance towards sponsorship with mobile services will become as natural as advertisements on radio and TV. The advantage, of course, is that in 3G the advertiser can avoid wasted contacts as the user is closely identified. The advertiser can be sure to reach the right target segment with the advertisement that is displayed. The more accurately the advertisement can be addressed, the more likely it is to be accepted and even wanted.

8.7.2 Buy your ad on the top of search engines

Search engines have introduced yet another way to get extra money from advertising. Search engines on the fixed Internet, such as Google, display advertisements next to the search results. This practise has not affected the popularity of using search engines while many users have found considerable added utility from seeing an advertisement which is very topical to the search. Again, the line between information desired and unwanted spam can be very thin.

8.7.3 Cross selling (own products)

Fixed Internet portal managers declare that the most effective sales person of a company's products is their own portal. This is likely to be even more so with 3G portals where you are expected to pay for a substantial part of the content. Spending and paying for services will be much easier in 3G due to the natural built-in billing capabilities of the mobile operator. The charging and billing issues are discussed in their own chapters (14 and 15) later in this book.

8.7.4 Customer loyalty programmes/clubs

Consumers are often attraced to customer loyalty programmes such as the frequent flier clubs of airlines. The portal in itself becomes an important sales channel for the company's own or a third party's products, which can be enhanced by 'mobile score' programs or 'mobile customer clubs'. Typical of loyalty programmes are those where the consumer collects bonus points that can be exchanged for rewards. The more a customer spends via the portal, the more bonuses are offered. For the operator and partners the benefit is that the more the customer spends, the more information about interests and consumption habits are generated and can be registered, even sold.

8.7.5 m-Commerce (partner marketing)

3G offers excellent opportunities for m-commerce. Quite distinct from e-commerce on the fixed Internet, the mobile Internet allows for accurate authentification of the customer. The 3G environment also provides a natural money-collection and tracking system from the mobile operator's billing system.

8.7.6 Multi-access portal

An evolutionary step in the digital convergence is the multi-access portal. Increasingly with the fixed Internet, mobile Internet, digital TV, etc., access to elements and sectors of the digital services universe, there is a need for one portal to handle access to all. That portal could be replicated on all devices maintaining the same look and feel, and give access to each network. Where the mobile operator is part of a larger ISP, fixed telecoms and broadcasting corporation, this may be relatively easy to accomplish. Otherwise it will require considerable amounts of integration and compromise through partnerships.

8.8 Closing the portal

This chapter has looked at the portal as it emerges as a new but significant tool for marketing in 3G. Many of the concepts we discuss are adapted from how portals have evolved on the fixed Internet, and we are likely to have missed significant aspects of where mobile portals will bring about totally new opportunities. As such, we are guided by Albert Einstein's thought guiding all involved in bringing about the new: 'Imagination is more important than knowledge.'

Marketing executives who stop advertising to save money are like people who stop the clock to save time.
— Paul Harvey

9

Promotion
Building Desire

This chapter looks at the promotion side of mobile services, from service awareness building to actual sales and purchase of services. As with many topics in this book, advertising and promotion of new mobile services are not well known to mobile operators, who are used only to selling handsets and subscriptions. Their experience is limited regarding what is needed to market their services. Promotion in many marketing texts and theories is considered to include sales promotion and sales. For this volume, we are placing sales as a separate item and it will be dealt with in a subsequent chapter (Chapter 13). For a simple way of dividing the issues, sales work is intended to get the customer to make the purchase at *that time*, all other activities to raise the prospective customer's interest and awareness and leading to sales, is promotion.

We look at promotion from the angles of advertising, publicity and public relations, branding, and product (service) placement. We also examine viral marketing. In short, promotion is the way to your customers heart, or the method to conquer it...

9.1 Is the classic marketing mix all mixed up in 3G?

To begin, with it is important to distinguish between marketing and promotion. It is a common misconception to think of marketing as advertising or promotional methods. Yes, advertising and promotion are part of marketing, but marketing involves all activities carried out to result in the sale of products and services. Marketing is a much broader term than is promotion, as marketing includes service creation, pricing, distribution channel management, etc. The ultimate aim of all marketing activities, including promotion, is to result in sales of services and products. Even in 3G the basic principles of marketing hold true. However, due to the revolutionary convergence of different media, innovative marketing mix strategies are made possible. In the following sections we will dig further into these new opportunities.

9.1.1 The AIDA rule

Promotion always should follow the AIDA rule, i.e. create Attention (service is perceived) and Interest (service seems attractive) which hopefully leads to Desire (to buy the service) and finally Action (actual acquisition).

Attention is achieved by delivering information. Information is conveyed by numerous means including the press, by promotional activities such as advertising, or simple word of mouth. In this first phase of creating attention, strangely enough, it does not matter that much what kind of service evaluation the message contains — the initial attention needs to be raised. Some companies go as far as to use annoying advertising explicitly to get initial attention. The marketing plan needs to cover the attention gathering method. Often with good press relations the attention factor can be taken care of.

A good PR plan will be a strong competitive edge in many other areas of 3G as we will see in the following part.

Interest goes hand in hand with attention and both interest and attention spans of customers are getting increasingly shorter. There are several pitfalls in 3G interest creation. A common mistake is to focus on the technology and various acronyms that are prevalent in telecoms. The average user is not interested in UMTS, WCDMA, IMT-2000 or 3G and rapidly loses interest if the message is too heavy on technology. Another common mistake is to disappoint users by creating too much hype in the initial stages of service launches whilst expectations are not met. This will radically elevate the threshold for interest creation for other follow-up products in the process. The questions relating to product creation are described later in this book in the chapter on product development (Chapter 4).

Turning interest into *desire* is like getting from a customers's brain (rational) to his heart (emotion). In this stage — described in detail in its own section later — branding appears. 3G has often been criticized as being too technology driven; and new technology as such is a very weak sales argument for the general public.

Even the strongest desire to buy is useless if it does not result in *action*, i.e. the buying process itself. Many marketing factors play a role in this phase such as pricing, placement (distribution channel and sales) and even mere availability of the service. Once again we emphasize an appropriate pricing level for 3G. Tariffing is discussed in its own chapter later in this book, but the rule of thumb is to aim below the pain threshold. 3G services are mostly mass-market services and their adoption would be squashed if prices are too high.

As 3G services are mostly quite abstract in nature, some services are remarkably intangible and marketing them will pose considerable challenges. Inspite of this, it should be made very easy for customers to acquire and consume 3G services. This means rethinking channel strategies and partnerships.

9.2 Crossing the 3G chasm

Geoffrey Moore's 'Crossing the Chasm' theory is a widely accepted model of the life cycle of high technology adoption. Moore identified the chasm as the critical point in the adoption of high technology products and services, which determines whether the mass market will adopt the product or service in question. The technology enthusiasts and visionaries are quick to try almost any new innovation or invention. It is the early adopting pragmatists who need to be sold on the benefits of the new technology for its mass market to become possible. As we can see in Figure 9.1, the main challenge is to bridge the divide between the innovators in the early market to the followers in the mature mainstream market.

Moore's theory applies to 3G with the some variations: first there is less time to establish the early benefits of 3G. The market has been burned by over hype relating to WAP and so consumers, industry analysts and the media are suspicious of whether 3G can provide any of the various expectations that have been assigned to it. The market will also witness global players entering most markets almost simultaneously. This has never happened before with any other comparable new technology. With personal computers, the USA was the early adopting market. With 2G cellular telecoms it was Western Europe, and Scandinavia in particular was the early adopting market. With home electronics and personal electronic gadgets Japan has been the early adopting market. With 3G several networks went live within a few months, and by the spring of 2003 a dozen 3G networks had gone into commercial production on four continents. This global simultaneous launch of a major new technology is unprecedented.

The simultaneous launches around the world involve not only the global and regional mobile operators, but also the equipment vendors. As 3G, and in particular the predominant W-CDMA standard, provide an unprecedented uniform standard for user equipment, more manufacturers see opportunities from handset sales to PC modems and custom 3G terminals than ever before.

Figure 9.1 The challenge is to 'cross the chasm' to the mainstream market (theory by Geoffrey Moore)

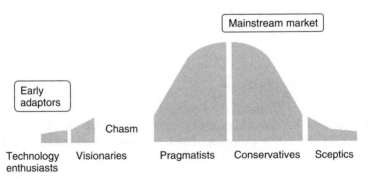

The rewards from mass market economies of scale bring added interest. For all manufacturers who intend to be winners in 3G, the race for early preferred terminals can mean the difference between success and an early exit from the market. For these reasons, we believe that the chasm will be more decisive but of even shorter duration than with most other technologies. Similarly, the expected global growth of the 3G terminals and services market suggests that the growth phase of the mainstream market is greater and the growth curve is steeper (the tornado resembles more a hurricane).

The chasm in 3G shows following potential pitfalls:

Pitfalls of chasm in 3G
- Pricing as cornerstone;
- Ability of different parties, i.e. competitors, to cooperate;
- Ease of use;
- Over-hype and disappointment;
- Real value of service for the customer.

2G services have shown that word-of-mouth should not be underestimated, and this applies to 3G, too. The viral marketing effects of word-of-mouth with early adopters showing the exciting new service to their friends and thus helping to recruit them has been an increasingly important element of mobile services. Some services which at first seemed to be very limited potential niche services — like SMS was for many years in 2G — may emerge suddenly as being very popular when the momentum is achieved in the viral marketing effects.

9.3 Public relations and press relations

Public relations can be a relatively inexpensive way of making your company and services known. In 3G the PR strategy is under careful consideration, since companies do not have money to waste. This, of course, does not mean that PR is for free, but it is definitely less cost intensive than most other marketing activities. The hype around the new technology in itself is attracting press attention so PR is easier in this 'sexy' business. A journalist's work consists of delivering news, and 3G services providers do have increasingly interesting and exciting news items — it is just about bringing the two together.

Journalists are so called message multipliers, i.e. they will spread information to large numbers of people and thus to numerous potential customers.

In contrast to marketing articles, press communications enjoy credibility and the image of being independent that a company's own advertising never achieves on its own.

The in-house communications department should have enough resources to deal with big launches and have excellent press contacts. However, the use of a high quality PR agency is highly recommended. Additionally, for optimal effect, PR and marketing will need to be synchronized as far as the tone, the timing and the message itself are concerned.

Regardless of what level of realism or hype happens to exist when you are reading this book, it is good to remember that public opinion, like economic cycles, tends to assume wave-like shapes. First, 3G was unanimously hailed to be the next killer technology in itself and touted as a 'can't miss' opportunity. Next the despair set in and all joined in the chorus to condemn 3G to certain doom due to high investment costs, etc. A good press plan will counteract the anticipated the wave pattern and adjust accordingly, to balance against times of too negative reporting, as well as dampening the level of hype.

When planning the launch of a a new service, a typical PR action plan would consist of a press release, press conference, product demonstration and product information material. Since the innovative air of 3G allows more unorthodox communication, this can be used to exploit the possibilities of 3G. A launch event, for example, can be used to provide a press relations forum for, say, one of the selected major stakeholders. The important thing to keep in mind with press events is to be absolutely sure to meet expectations. The only thing worse than disappointing your customers is disappointing the press. Bad publicity can even turn a satisfied customer into an unsatisfied one.

Be sure that when — or even before — the first services are launched the VIP press contacts are among the first to get to use them. The press release can be tied to the services: information should be provided on the fixed Internet and of course also via the mobile portal. Why not provide VIP guests and the most influential journalists with a 3G terminal and the instructions to switch it on at a certain time in order to broadcast the press conference and demo your 3G technology directly.

9.4 Advertising mobile services

Advertising is paid promotion to raise the interest or awareness of one's product, service, company or brand. The most common advertising media are TV, radio, newspapers, magazines and billboards. The fixed Internet has recently been growing as a new advertising medium and the mobile

Internet — initially SMS but already also WAP — is seen as the latest new frontier. With the general trend of content and delivery convergence, the line between advertising and content is becoming blurred. Various 'infomercials' exist on TV as paid advertisement programmes. The Yellow Pages and used-car ad papers are providing advertising as content.

When mobile operators were selling only voice services, advertising was relatively simple. The service (voice calls to and from mobile phones) was only available on mobile networks, the branding, service proposition, competitive argumentation, etc., were relatively clear. With the new, advanced, mobile-services opportunities of 3G and 2.5G, the various parties with an interest in promoting any given service may have differing, even conflicting needs. For most service providers, the mobile Internet is just one channel among many. The mobile operator may see many of its partners advertise the mobile Internet option as only one — even just a marginal — aspect of their service delivery options.

Whenever advertising a tangible product, such as a new model of mobile phone, the advertising message is usually rather simple. Describe what needs the new gadget is intended to serve, how it may differ from its competitors and create an interest in buying. With services as abstract as new mobile services, the actual benefit may be much more difficult to identify and communicate effectively. One of the big benefits of mobile services is the Moment attribute — getting fast access to sudden needs of information. That is still clear, but trying to sell the idea of security in various possible sudden emergencies — the car breaks down, the airplane is delayed, etc., approaches the challenges of selling insurance. Few insurance advertisements reach the excitement level of most mobile phone ads. The same is true on

Table 9.1 UK consumer preferred sources for news, ranked in order of preference

(1) Newspaper
(2) TV broadcast news
(3) Radio news
(4) Cable TV / satellite 24-hour TV news
(5) SMS mobile news
(6) Internet

Note: SMS news are already used by 9% of the UK consumers in 2003
Source: UK Independent Television Commission October 2003

the Money attribute, it is hardly a reason on its own to sign up to a 3G service so that you can pay your monthly electricity bill via your mobile phone.

The operator and his content provider and application developer partners need to find respective roles and synergies, and to decide who promotes what and through what means. In the above example of the electricity bill, it may not make exciting TV ads, but it might be very worthwhile for the electicity company to promote this new payment method, for example, in its customer contacts, services centres and marketing literature.

The existing advertising channels have rather established pros and cons. TV advertising is costly, reaches mass audiences only at very general demographic divisions, can be very effective, and takes time to produce. Radio advertising is usually faster but reaches much more targeted audiences and is usually cheaper. Newspaper advertising covers general audiences and costs in advertising vary greatly with circulation, size and placement of ads, etc. Magazine advertising tends to be most targeted but production times are often lengthy. Billboards offer yet another type of advertising, ranging from large signs along motorways to poster size ads at bus stops, etc.

Internet advertising promised radical improvements on older advertising media. The information carried on the network allowed general targeting, such as noticing that someone connects to an American website from an Italian location and showing banner advertisements with the text in Italian rather than in English. Of course context-based ads could be placed as in targeted magazines; for example, if you visit a website for tourist information,

Figure 9.2 Effectiveness of SMS advertising campaigns, based on a survey of six campaigns run in the UK, Germany and Italy (Interquest March 2002)

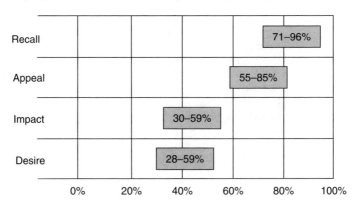

you might see ads for airline specials to your destination. Web browser technology brought the ability to place cookies onto personal computers to identify the users. Banner advertisements could then be used to target ads even more precisely. Banner ads introduced a new level of advertising target audience activation — viewers could click through to the website of the company whose banner was being shown. The big problem with banner ads was that after the initial novelty factor wore off, the click-through rates kept dropping.

As with all marketing, advertising operators and their partners should set clear objectives of what they want to achieve. Advertising can be used to generate direct action by prospective customers, such as redeeming a coupon, visiting a store, answering a poll, participating in a drawing for prizes, etc.

9.5 Publicity

Words that that can easily get confused with each other, publicity, PR and sponsorship are distinct aspects of promotion. Publicity is any visibility in the public domain, usually covered by the media. Publicity can be planned and unplanned, positive and negative. For example, if a fire breaks out at a company office and that gets visibility in the press, it gets publicity. There was no payment for the media coverage which is why it is not advertising. Companies are eager to attract positive publicity as the general public trusts news stories much more than advertising, even if they deliver the same message. That is why companies are eager to attract (positive) publicity. Quoting Oscar Wilde — the only thing worse than being talked about, is not being talked about — has given rise to the theory that there is no bad publicity. Even bad press keeps the company in the media interest and the public's eye. With recent catastrophic prolonged bad publicity news stories involving Enron, Arthur Andersen and Worldcom, however, we are of the opinion that there is such a thing as too much bad publicity.

PR, or public relations, is the proactive behaviour to try to get media attention and coverage. PR attempts to achieve publicity. Companies employ PR experts in-house and hire PR agencies to support them in achieving media visibility in an increasingly competitive PR environment. PR campaigns can be mainstream press events, with invited press to cover, for example, the release of a mobile operator's annual report. At the other extreme, PR events can be most creative, such as the spectacular launches of each season's latest Formula One cars. The aim is to get the press to become interested in the company's current story and to report on it.

Publicity and public relations (PR) usually involve more creative work and there are typically no set fee structures as there are in advertising.

A publicity campaign usually intends to receive press coverage or visibility at some mass audience event. Publicity also typically results from the daily running of large businesses, and the achieved publicity can be both good or bad. Publicity rarely is intended to achieve immediate action by customers.

9.6 Sponsorship and product placement

Sponsorship is the activity of providing funds or other support to some activity, usually with a related visibility aspect. Sponsorship is a dramatically growing area of promotion and ever more creative ways are found to promote. Traditional ways of sponsorship include banners showing company logos at events and locations that are not considered very commercial, such as mass-audience sporting events from football to horse racing to the Olympics. Sponsoring company logos can be inserted into the literature involved with the event or activity, such as brochures, programmes, etc. In very traditional areas such as museums, classical-music concert series and the opera and ballet, a sponsoring company might only get a one-line mention in the back of a catalogue with 50 other sponsoring organizations, companies and individuals. In other cases it may involve the company logo blatantly pasted on the very individuals involved, such as the sponsor logos sewn onto the overalls of racing car drivers, the jerseys of football players and the helmets of ice hockey players, etc. The visibility received through sponsorship can be exceptional if the event or occasion is suitable. Perhaps the most extravagant example was the sponsorship — and related advertising — of a recent Russian space-rocket launch.

A special form of sponsorship and/or advertising is product placement. Typical examples are the various products that are placed within a movie, such as James Bond drinking Bollinger champagne or driving an Aston Martin car in his movies. Product placement has received an increasing role in promoting global brands with soft drinks, beer brands, watches, clothing, etc., of TV shows, movies and now even video games, being a new way to promote products and services. Competition can be fierce between rival products or brands, such as Ericsson and Nokia both competing to be the mobile phone of choice for the James Bond movies. Naturally with the selection of any product placement opportunity, the promoting company will need to decide whether the intended target audience is appropriate, and whether that project is a cost-effective means to achieve the promotional goals.

Sponsorship and its related product placement are a type of hidden advertising. The success of any publicity, PR or sponsorship campaign is dependent on a wide variety of factors starting from the relative visibility of the campaign and any related events — a sporting event moved to a less favourable day because of bad weather for example.

9.6.1 Viral marketing and communities

A very powerful new method of promotion is viral marketing. With viral marketing the initial message is given to the first contacts, with the intention that these first contacts then spread the message to their friends, colleagues, etc. Such campaigns can be very successful if targeted well. For example, a 10% discount coupon of Eminem's latest music CD could spread very fast if registered Eminem fans were first to receive the coupons, and then given the opportunity to spread the coupon via the mobile network to friends. The initial contacts would of course mainly only forward the ads to other known fans of Eminem's rap music, so this method would achieve very strong targeting, if the initial contacts were found with enough precision.

9.7. Conclusion

Promotion is how the awareness, interest and 'pull' for a service is created. Promotion helps sales by introducing sales arguments. Promotion serves to build brands and is used also to reinforce purchase decisions. While marketing is much more than promotion, much of the early visible work in marketing is promotion. And do not forget the effects of viral marketing, it is as James A. Robertson said, 'The most powerful words in marketing are "Watch this!"'

The empires of the future are the empires of the mind.
— Sir Winston Churchill

10

Branding

Diamonds are a Girl's Best Friend

A 'brand' is something that makes Coca Cola taste so much better than any other soft drink — it is something that somehow appeals to us, makes us believe we get better service, better quality and finally improves our quality of living. How is it possible? How does it do that?

This chapter looks at branding. We discuss what a brand is, what is its impact on the business, and how brands are built, developed and expanded. For the mobile Internet and 3G uses, co-branding is likely to

---

3G Marketing: Communities and Strategic Partnerships   Tomi T. Ahonen, Timo Kasper and Sara Melkko
© 2004 John Wiley & Sons, Ltd   ISBN: 0-470-85100-7

become a very significant issue, but before looking at the mobile telecoms issues relating to it, let us start by defining what a brand is.

10.1 What is a brand?

In a nutshell, branding is attaching certain associations and emotions to a company or some product or service. This is particularly important in markets with very homogeneous products or with abstract, intangible services like those delivered via a mobile phone. The trick is to trigger a reaction in the customer's (or potential customer's) brain or — more accurately — in the heart. A good brand runs a certain programme within us. By connecting feelings and emotions to a brand, one achieves the desired association every time the customer is in touch with the brand.

Why does somebody choose the laundry detergent Ariel over and over again? It is not the technical superiority that makes us prefer one product to another, but the programmed belief that this particular washing powder makes our laundry whiter, preserves the colours better, or makes the shirts softer.

A brand does nothing to us, we do it to ourselves. Although a brand is usually built around a symbol or a name, the brand releases certain associations and emotions that we have grown to attach to that brand. A brand exists in our minds. It's quite funny that most of the *value* of some world famous stock listed companies — for example, the Coca Cola company, — is not necessarily in the processes, production methods, or patents, nor in the tangible assets or any other 'balance sheet items'. The greatest asset may actually be in the minds of people. It is in the notion of how people around the world perceive the soft drink, and most of all how and to what they associate themselves with when consuming the beverage.

10.2 Why brand?

Brands help consumers to navigate the world of consumerism. We are faced with remarkably numerous choices on a daily basis, too many in fact to be able to function unless we had preferences through brands. Which toothpaste, which morning cereal, which margarine, which orange juice, etc. If every time we had to make an active conscious selection process, we would not make it out of our homes in the morning.

Table 10.1 Global ranking of brands 2003

(1) Coca Cola
(2) Microsoft
(3) IBM
(4) GE
(5) Intel
(6) Nokia
(7) Disney
(8) McDonald's
(9) Marlboro
(10) Mercedes

Source: Interbrand 2003

In a shopping situation the clutter of choice is even worse. Imagine yourself going into a shopping mall to look for new tennis shoes. In the sports shop you will find a huge wall of nice looking tennis shoes. How do you know which ones to even try on? The shopkeeper has made it easy for you, he/she has put one rack full of Adidas shoes, another full of Nike and a third one with Reebok, not to mention four remaining other racks. What do you do? You probably walk directly to one of the racks (read 'brands') where you think you'll find tennis shoes that will suit you best. Maybe you had a pair of that brand earlier and they were just great, maybe your brother-in-law disliked one brand and you will certainly not make the same mistake. In a hurry, you probably don't even think about it, one rack just appeals to you more than the others.

10.2.1 Brands aid in decision

One of the main purposes of a 'brand' is to make it easier for the consumer to make their decision/selection. A brand, when experienced in a positive sense, increases customer loyalty quite substantially. In some cases users can become fanatical about brands such as Levi jeans, Sony Walkman, etc. Therefore, it is important to acknowledge that a brand is *a promise to a customer*. It is a promise to supply the same positive emotions and experiences that the person experienced or associated with earlier. Sometimes a brand is used as an expression and a symbol uniting a group of people — a sort of community sign. Some biker clubs are only happy with and accept members who ride motor cycles of one particular make — Harley Davidson.

10.2.2 Brands and teenagers

Brands are important to all ages of consumer, but particularly to teenagers. Teenagers enter the consuming world with their own personal funds, usually allowances from their parents, and increasingly also demand to make the choices on purchases that parents make on their behalf. Children are particularly strongly affected by peer pressure. In school they do not want to be singled out as 'not belonging' by being associated with the wrong type of clothes and other gear. Most teenagers feel insecure and have low self-esteem, that is why they try not to stand out from the crowd. In order to achieve this they want to dress up like all the others — jeans and T-shirts, the right brand of sneakers and RayBan sun glasses — you name it.

10.2.3 Brands and price

The more a class of products is homogeneous, the more any product within that class needs branding. Where products that can easily be substituted by others it is increasingly difficult to differentiate and get customers to pay a higher price. Yet it can be done, even in products seemingly identical, like water where strong brands can hold remarkable market share and command higher prices like Evian and Perrier. These brands are strong sellers even in markets where the tap water is pure and clear, such as Switzerland, Sweden, Canada, etc.

A brand is a way to differentiate one product from others similar to it, not necessarily in technical or other scientific ways, but in terms of the image inside the minds of people. The advertising for most consumer goods with strong brand association — like tennis shoes — is not very informative. The advertising is often more like a mini movie, bursting with energy and loaded with strong positive feelings and 'generally accepted values' that the brand owner wants the viewer to associate with their products. Think about cars, what makers would you associate with the following qualities:

- Safety;
- Reliability;
- Speed;
- Aristocratic.

We all know that a Volvo will engender a very different image to that of a Porsche. One of the goals of branding is to command a higher price for our products or services (the other main goal is customer loyalty). Branded goods have lower price elasticity than their non-branded substitutes. A low price elasticity means our service is less sensitive to changes in price. When increasing the price for our branded service, the ultimate purpose is to shift from a 'perfect competition' where all competitors are perceived as being equivalent, towards a 'monopoly' (other models of competition economists generally refer to are 'oligopoly' and 'monopolistic competition') where only one competitor can be considered. If we can achieve some demand in the minds of our target customers, so that they only consider our brand, then this monopolistic condition has been achieved. Then, as in any monopolistic market, the situation allows us to price out the goods independently and thus yield better margins from the markets.

10.2.4 Brand and loyalty

In addition to attempting to achieve a higher price level, the other main goal of branding is to increase customer loyalty. Branding enforces customer relations and loyalty. It makes a company unique and distinguished vis-à-vis its competitors. Branding is a strong success factor in gaining and maintaining market share, since technology-driven marketing does not deliver strong market success, and price competition does not provide the basis for long run competitiveness.

The return on investment of branding is considerable and can often be directly measurable. Given that a certain added value is perceived, the willingness to pay for this value rises. Maybe somebody would not pay double the price for a pair of Diesel jeans versus a no-name product, but there are enough who will. Mobile operators will need to learn from more mature markets and not give up. It is not at all impossible to think that in the future some mobile brands will achieve strong brand associations and can charge more for their services because of it. If it can be done for plain water and for 'worker pants' — as blue jeans originally were — it can also be done for telecoms services.

10.3 Needs to be comprehensive

Branding is not only a name, a logo, the claim, or the advertisement along with colours, message, music, feeling. This 'look and feel' is of course a central part of brand building, but there is much more to take into account. Whenever our target customer or prospect is in touch with the brand — no matter in what way — that person should be encountered with the same consistent message everywhere. Branding is much more than pure design: it is a comprehensive perception of things, such as the web site (content, navigation and, of course, appearance), the way a company addresses its customer, the (technical) support and all correspondence from buying a service up to billing and after sales. The calling centre will need to convey exactly the same image as the TV ads.

McDonald's corporation, for example, has established children's birthday parties as one of its brand-building vehicles and is training their customer service personnel at the counter to enforce the 'family restaurant' feeling in service. Amazon is advising you on what books you might like and has as easy to use online transaction processes. Branding is not about hypnosis, propaganda or false pretences, since there is real value in the way customers are treated or how the service is delivered.

10.3.1 Brands in mobile telecoms

Now that we understand what branding is, how does it apply in the mobile services area. The first point to keep in mind is that the brand starts with the handset, phone or terminal. In some parts of the world, the phone brand is known. Typically in Europe the mobile phone is bought as a Samsung, Nokia, or Motorola phone, etc. In the Americas the phones often are branded by the mobile operator (wireless carrier), i.e. the identical phone would be known in the USA as a Verizon, Sprint or Cingular phone for example. Parts of Asia follow the American model, such as Japan and Korea where manufacturers brands such as NEC, Panasonic and Samsung are hidden behind brands like NTT DoCoMo, J-Phone and SK Telecom. Other parts of Asia behave like Europe.

As the phone is switched on, and the display is activated, the mobile operator will be in control. From that point on, whatever is on the screen is controlled by the operator. Here is where the brands come to play. In the 3G world, as in all of the mobile Internet, the operator portal will come into play. We will be greeted by the welcoming screen of the portal that is activated. Then on the small screen we are likely to see a dance of logos of various services, content providers and software applications.

10.4 How to build a brand

Let us assume we have no brand at all, like a brand new 3G operator, or the brand we have is considered so negative that we find it damaging the market success of our service. How can we make our services as well known as a brand like Pepsi Cola (it has only been in the markets for about a hundred years, right)?

Brands are built and developed over many years of constant and consistent work, a globally recognized brand takes decades to build. It is not surprising, that we find old consumer products like Coca Cola, Heinz or Levi's among the best known brands. However, there are no short-cuts to building a brand. The brand and image can only be built on clear focus and consistent promotion of it.

10.4.1 Where do I begin?

Branding always starts with analytical reflection. At first, the brand-building process is a thorough look in the mirror. The questions to ask yourself and your interest groups are:

- what makes my company/product special compared to others?
- why do customers buy my service?
- what is my unique sales proposition?
- what 'character' my company has got?
- what are my strengths — weaknesses — opportunities — threats?
- how is the competition positioned?
- etc.

These prompts define the actual *status quo* to begin with. Then, the wanted and less wanted aspects have to be defined. Incumbent telcos still sometimes have a somewhat old-fashioned air sticking to some parts of their operations. Still it depends on what attributes you attach to which segments and in what way. Yet you cannot turn a mule into a thoroughbred. Fortunately there are still markets that need mules. However, if some target market demands donkeys you could try to transfer some donkey-attributes to your mule, since mules and donkeys are very close imagewise, but if you act as if you were a racehorse, you will not be credible and will fail.

Brand creation starts with recognition and a good concept. Nokia's brand building plan consisted in making the name known by the products and then connecting — besides people — more substance and emotion to the name. We dare say that among Nokia Mobile Phone's most crucial factors of success and brand building were self-evident features like ease of use and user-friendly design — not emotion-laden TV-commercials, even though, obviously, promotion has been important as well.

10.4.2 Employee buy-in

Since abstract (technological or online) brands have much to do with service and processes, internal communication is as important — if not more so — as external communication. If a company fails to inform and motivate its own staff, all the effort to create a brand is in vain. Personnel have to understand, learn, accept, and virtually adopt the brand message. The brand becomes their attitude, their work, their service, the way they think and commit to the company or its products. An employee has to realize: 'I AM the company brand or at least part of it.' This is only possible when the brand is seen as positive and fulfilling.

10.4.3 Damaging the brand

The spring of 2002 brought about several instances of major brands in big trouble, mostly due to accounting problems. With the collapse of Worldcom and Enron, the former accounting and auditing giant, Arthur Andersen lost its credibility and disappeared. While other companies in the past have lived through dramatic crises that have threatened the brand, such as Mercedes Benz with its introduction of the A series and the subsequent finding that the tall but short automobile might roll over, a company can usually survive such a crisis if its product offering is solid. The more a company's service is intangible, the more it is dependent on the brand. As Arthur Andersen was known for auditing and accounting *services*, when that service itself was called in doubt, the end was imminent.

The more a company spreads its brand beyond its core business values, the more it puts the brand at risk. Similarly every co-branding opportunity brings with it threats of damage through unanticipated activities by the partner.

10.5 Multiple brand messages

So far we've discussed what a single brand is. In reality, brands coexist with other brands, brands may work together in co-branded situations, and brands can be extended into new areas. Branding includes issues such as sub-branding. We will discuss these issues where the multiple-branding situations occur.

10.5.1 Cross branding

One brand building approach is to use an old existing brand and develop from that. This is called cross branding. A good example of this is Hollywood's merchandising activities. Most parents can tell of experiences with merchandise of an popular animated childrens movie especially if the film has been a blockbuster. At such times it is very common to bump in to the 'brand' of the movie in most peculiar places, from hamburger restaurants, kids clothing boutiques, candy store, ice skating shows, to collectors/trading cards, etc. Cross branding has become a true industry of its own. Many strong brands are doing this today from Marlboro and its lines of clothing and accessories to Jaguar, which sells a wide range of products from leather-covered sofas to umbrellas and keyholders under Jaguar name and brand. The Spanish car maker Seat — owned by the Volkswagen group in Germany[†] – marketed its new models a few years ago with a slogan 'System Porsche'. The engines were designed at the Porsche engineering centre. In the information technology sector it is quite common to see that some services are 'Powered by Oracle' or 'Powered by IBM'.

10.5.2 Sub-branding (overall company branding versus product trademarks)

Defining the 3G branding strategy is very much dependent on the overall marketing strategy of the operator. Some operators have gone a long way to build their brand identity and are rather reluctant to change the name of the company due to its history, even though to today's taste the company name might not be hip enough or gives too much of a hint to its telegraphy/telecom roots. Furthermore, most existing operators have customer bases so vast and diverse that a company brand that suits all its customers is nearly impossible to define or create. For example, the brand that projects reliability, safety and stability will appeal to older people and traditional

---

[†]The Volkswagen group owns among, other things, several other famous brands like Audi, Porsche, Lamborghini and Bentley.

business executives. Those same brand characteristics do not appeal to the youthful and innovative, who will see that brand as old-fashioned, slow and unexciting.

One extreme solution is to end up with a multiple brand personality where one brand attempts to be many different things to different target audiences. This is very hard to get right and to sustain over time. This strategy often delivers a wishy-washy image in the market. At the other extreme, the operator can decide to reduce its branding appearance to some characteristics only and risk therefore to alienate many existing customer groups.

This conflict can be overcome by creating a family of brands or trademarks. The company name serves as a combining factor for the different products, but is not underlined in a special way. The products or product groups, however, are assigned their very own personality directly tailored to appeal the target segments. The product brand is thus more definite, carved out, distinct and powerful. A good example can be drawn from the automobile industry where a manufacturer like Ford will build various brands around its family of cars such as Jaguar, Volvo, Lincoln, etc. 3M — though having a good reputation and known trademarks with different products in various business sectors — tried to bring up the name of the company itself. This worked out well by presenting the success story products of these different sectors and creating a special '3M technology excellence' feeling. Even though 3M's portfolio is highly diversified, the company tried to create some kind of common feeling.

It is clear that the creating of the product brand has to be in line with the company brand in order for them not to collide and damage one another. Of course, sometimes the company itself has no special brand at all and is reduced to a mere business name without its own personaliy. This is often the case when a company has diverged to many different business areas and holds strong trademark names (brands) in its portfolio. A customer does not necessarily mind if his favourite toothpaste is Procter and Gamble or Unilever — as long as his heart says that this very toothpaste gives him the best protection, smile and breath. People would be buying Pepsodent without even knowing who the brand owner is.

10.5.3 Co-branding

Product trademarks and brands become very significant when cooperating with partners. If a partner is in a very strong and important position, not only must the owner company's requirements towards the brand's character be taken into account, but so must the partner's. This is, of course, especially important when marketing or sales activites are executed together. Co-branding with new partners is easier and less risky with a new, common product brand than if the company name were at stake.

However, the amount of sub-brands must not be too high. It is too costly and confusing for both in-house staff and the target markets to try to keep up with and communicate too many (maybe even disharmonic) brands within the company. Concentrate your energy is a maxim that applies for many things in life — also brands. Bear in mind that sub-brands could be an option as we mentioned. Sub-brands can be very effective when a new product or product line is too different from the company's own major brand. It just has to be made sure that the sub-brand does not damage or weaken the main brand's standing.

10.5.4 On-line branding

The heart of online branding is the web site. Run this test by two people — a girl of about 13 years of age and an elderly woman of about 60 years of age. Make sure neither of them is particularly interested or knowledgable about telecoms. Do not consider these as practical questions for your web master, your brand manager or yourself. Ask your quick survey team to consider these factors and tell you what they thought:

Website quick test
- Are you attracted to the site?
- Are you confused by what it communicates?
- Do you want to stay a while or do you have the urge to surf on to another site?
- Do you understand (in 30 seconds) what the business is and whom the company wants to address?
- Is there interesting information on the home page?
- Is navigation easy and quick?

The quick test will reveal if your company has found the thin line between too much and too little. It also helps to reveal whether or not your company can communicate to diverse audiences. Of course, there is much more to it than that. Online branding is not only about flashy banners and the accurate colours on your website. Branding on the web — as traditional branding — defines itself via processes in the first place. How quickly do I find the accurate information? Do I find the correct contact information for my material? How can I give feedback and how fast will I get an answer in the way I wish to be contacted (this could be by e-mail, a phone call, a letter, an SMS text message, etc...). Is the site functional yet entertaining? Visiting the site should pay off in some way or another. The best way to develop your service and its brand is to ask your customers and collect their ideas and feedback. Make sure that you thank each customer who takes the trouble to give you ideas and make sure you inform them when you've made some activities in response to their idea.

The more business processes are rolled out online, the more important a company strategy and the 'how we do it' gets. The most essential fact to keep in mind is: keep it as simple and straightforward as possible. Online branding does not have the luxury of time.

A relatively low-cost online marketing method is that of distributing newsletters online. Subscribing (and desubscribing) can be done on the web site. The content of the newsletter is important and must be made worthwhile for the target audience to read. Try to ensure that readers always find something of concrete value in your mailings. No one wants to be spammed with unnecessary marketing. Again, the line between marketing chit-chat and customer care and information is thin. It also varies considerably across cultures and between countries. The word that describes online branding the best is user experience. Brands are about values and feelings. What could be better for a good feeling about a brand than good service.

10.6 Action plan for branding

Now that we know what branding is we will look briefly at some of the main items to be kept in mind during the actual brand management activities. While branding is a key part of the overall marketing mix, it is its own discipline and much has been written about it over the past decade or so. There is not very much specific to mobile telecoms or 3G in branding and we therefore suggest reading some of the recent branding books for more information. To end this chapter we will just offer a brief overview of some of the 'do's and don'ts' of brand building, as well as giving a brief outline for a plan.

10.6.1 Branding 'do's'

Some of the basic items to consider when creating a brand are:

- What is the company's strategy? (mission, vision)
- What is the company's positioning? (unique sales proposition, vis-à-vis competitors)
- What is the brand's intended personality and identity? Define the different branding aspects brand-as-person, brand-as-organization, brand-as-symbol, brand-as-product. Use associations to clarify such as: 'What sport would our company be?' Mountain climbing? Basketball? or 'What car symbolizes best our products and services?' A Volkswagen Beetle? Pickup truck? Jaguar? etc.
- What is the value proposition statement? How is the message conveyed?
- What is the tone of voice, look and feel of the brand identity?
- Is the brand differentiated enough from competitors, is it unique; would my target audience prefer my brand proposition to that of my competitors' offerings?
- Are there guidelines and information available to everybody in the company on how the brand applies to their work?
- Does everyone in the organization know and — more importantly — really *embody* the brand statement and message?
- Is the brand durable over time? What plans are there for building the subsequent work and delivering consistent messages over time

Branding is a task for corporate management. Naturally, the marketing director and marketing department are closely linked in the process, usually driving the branding initiatives. However, the range of actions that are necessary successfully to develop a sustainable brand image throughout an organization usually exeeds the powers of the marketing department. This is why the execution and whole mission of the brand creation process should be fully supported by and actively embodied by top management. Furthermore, there should be one designated person — a brand manager — who is in charge of the creation, maintenance and consistency of the brand. The brand manager monitors how well the brand is perceived in the market and makes proposals for activities to strengthen or defend the brand as necessary.

10.6.2 Branding 'don'ts'

The biggest mistake a company can make in its brand building is to have a conflict between the brand promise and reality. For example, a modestly priced car should not promise the performance of a Ferrari and the luxury of a Rolls Royce. In most industries with concrete tangible products, this kind of overpromise is rarely seen, but with intangible brands very serious overpromises or mispromises can be made where the brand promise collides with the reality of the actual service. The bad will that is generated by a disappointed customer in such cases is immense. Now, we are not talking about eventual flaws that will inevitably happen with any business as human nature, change and technology all occasionally conspire to cause failures or errors. We are talking about a fundamental, inherent, permanent and obvious mismatch between a company's communication about its brand promise and the actual delivery of its performance.

Another common mistake is to reduce the brand building efforts to mere marketing slogans or visual design of a new logo. Imagine the 'happy and friendly' amusement park whose employees are constantly discontented with their poor pay, unsafe working places, bad atmosphere among workers and the way they are treated by the management. Don't you think this would not eventually show up in the service? When there is an inherent conflict between the promise and the reality, customers will observe and of course will make their final judgement on the real performance, not the brand promise. The brand may be irretrievably damaged if such conflict is prolonged.

There might be counter-reactions, too. If the brand philosophy is forced upon the staff, the launch will end up in a total Waterloo defeat. Nobody wants to be brainwashed and moulded into a stereotype, with everbody wearing the same ideological uniforms.

Companies often tend to forget their most important people: the customers. Sometimes companies may go too far in focusing on the internal communication and neglect the customer. Brands, organizations and processes may be established too much on internal rather than customer needs. Again, a balance is needed. A certain part of the branding budget has to be invested in market research to keep track of brand perfomance. A good, well analysed foundation will pay back in the end.

10.6.3 Brand development plan outline

Phase 1: Internal view

- interviews of the management
- interviews with staff
- history – objectives
- products

Phase 2: External view

- competitors (competitor analysis)
- customers (segmentation)

Phase 3: Development of brand strategy

- target customer segments
- positioning
- set measurable objectives
- build marketing plan
- define communication plan

Phase 4: Implementation

- corporate design
- material
- corporate identity

Phase 5: Rollout

- commitment (has to be started in phase 1)
- company-wide comprise

Phase 6: Review and Refinement

- measure performance against set goals
- improve

10.6.4 Brands grow too

Even brands grow with time. 'Growing' here means changing and altering, but doesn't this contradict to the credo of 'consistency of the brand'? The trick is that the moves and changes to the brand image need to be delicate and subtle over longer periods of time. Many companies may adjust their logo by changing its font or the graphic element a little bit every now and then. Normally, a decade is a good time frame for traditional companies critically to evaluate their brands. For the information technlogy business the cycle is of course shorter but still measured in years not months.

Even the big, mighty and successful ones go through slight adjustments, although you have to admit that the Coca Cola logo still is almost the same that it was over a hundred years ago. Sometimes history even repeats itself: e.g. Coca Cola got back to a Fifties bottle design. Seeing a logo change through decades you might not notice the finesse of change from one iteration to another, but if you compare the starting point with the *status quo*, the difference might be bigger.

As a new phenomenon, communities in the connected age are learning to pool their power and initiate change. First examples range from political activism, such as the deposing of the ruler of the Philippines in 2001, to the word of mouth with viewers of the movie *Charlies Angels 2*. Communities of users can rapidly share information and take action — or inaction — and form a counterbalance to the power of branding. The herd behaviour of 'smart mobs' is very new but powerful development and one that mobile operators will need to study and understand. In the future, the role of communities will be ever stronger and branding must understand this and work with and through communities, not against them.

10.6.5 After the brand, what is left?

This chapter looked at how brands are used in marketing. Brands were considered as tools to increase the perceived value to the customer and to build an identity and create loyalty. Brands were looked at independently and in context of other brands within the 3G space. For many who are reading this book, the branding work is just at its beginning. For those it is good to remember what the author John LeCarre said about his work: 'Writing is like walking in a deserted street. Out of the dust in the street you make a mud pie'.

Every new opinion, at its starting, is precisely a minority of one.

— Thomas Carlyle

11

Service Adoption

Early Adopters are Different from You and Me

For any new technology there has to be a launch and the adoption of the technology will follow predictable patterns. Recently plenty of research has proved that there is a definite adoption pattern and, furthermore, that pattern is dramatically influenced by how so-called early adopters embrace the new technology. This chapter looks at the general patterns of how services are adopted and discusses theories about saturation levels. Then an analysis of the

---

3G Marketing: Communities and Strategic Partnerships   Tomi T. Ahonen, Timo Kasper and Sara Melkko
© 2004 John Wiley & Sons, Ltd   ISBN: 0-470-85100-7

business customer versus residential customer debate is covered. Finally this chapter identifies the needs of the group of customers called 'early Adopters'.

11.1 S-curves

It has recently been proved that new technologies are adopted in a typical, predicable pattern. These patterns were described by Rogers and Shoemaker (1971) in the Communication of Innovations. First users are called *innovators*, and statistically they have been identified to be 2.7% of the total population. Next come the *early adopters*, who amount to 13.5% of the population. Often, especially in telecoms and IT, these two are combined and called *early adopters*, in which case the proportion of the population is 16.2%. For the purposes of this book we will combine both innovators and the early adopters and use the combined term early adopters to describe all the approximately 16% who are first to adopt a new technology.

The next groups to adopt a service are called *early majority, late majority* and *laggards*. The early majority bring the usage to about 50% of the total user population, the late majority adds about another 34%, and the last group, laggards account for the final 16% or so.

When the adoption rate is plotted over time a curve emerges, which is commonly called an S-curve (Figure 11.1). Apart from the shape of the

Figure 11.1 How adoption and penetration relate

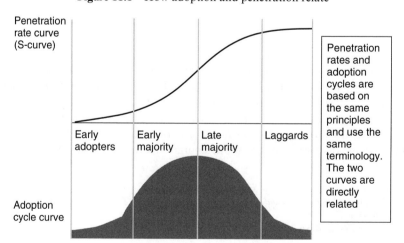

curve following the letter S, the other relevant issues with the new service adoption S-curve are the slope of the early adoption stage and the saturation point.

11.2 Where is the saturation level?

At one end we have the launch and the early adopters. At the other end we have *saturation*. The first issue about saturation that needs to be discussed is *what* do we measure. Penetration may seem simple — who has a (3G) mobile phone. We could measure the number of mobile phones. When all are on one generation and type of technology, it is a relatively fair measure, but when several new technologies are introduced in succession, and older mobile phones become obsolescent and obsolete, just counting the worldwide population of mobile phones would yield an inflated number. Much like ignoring the early generations of personal computers, which have been assigned to the scrap heap, so too we should not be counting older and unused mobile phones.

If not phones, then subscribers? This seems like a reasonable approach, and for a long time the mobile telecoms industry has been reporting its subscriber numbers and these used to correlate well with overall penetrations, that is until two new phenomena appeared. First the possibility of getting very inexpensive prepaid subscriptions produced a new crop of subscribers who might use a phone for a very short time and then abandon its use. Prepaid cellular accounts were invented only a few years ago in Italy and Portugal as a response to local tax laws. As the prepaid account system enabled a rapid expansion of the customer base to a younger generations of users, the innovation soon spread around the world.

The other phenomenon that changed the counting of subscribers was the surprising development of users having multiple phones. This was something that traditional telecoms forecasts and data could not predict. At home a family tended to have one phone, perhaps got another for the eldest of its teenagers, and perhaps even got a line for the Internet, but two lines for the same person to use voice was unheard-of. Yet dual mobile phone subscriptions and, increasingly, even triple subscriptions for the same person are not unusual in any developed mobile telecoms market.

The reasons for having multiple subscriptions are many. Perhaps the most common reason for multiple subscriptions is that one is the company phone and the other is a personal account. Another common reason is to switch between mobile operators and their fee structures. Yet another reason is for tax purposes, as in some cases if the employee uses the company phone for personal use, it is taxed as a benefit, but if used only at work it is considered a tool of the employer. A further reason would still be work- and home-related in those cases where the work phone number is widely distributed and phone calls might arrive at inconventient times. The user would get a personal subscription so that the work phone can be turned off for evenings and weekends, etc.

Still other reasons include network coverage issues and personal private reasons, such as having a private business on the side as a hobby or to supplement income. A second (or third) phone could be for more clandestine reasons such as to interact with a secret lover or to conduct illegal activities, avoiding taxes, etc. A widely reported study by Motorola suggests that up to 10% of all mobile phones are used in this way. There are very many reasons why people have two subscriptions, and some heavy users are starting to have three.

11.2.1 TV set analogy

Early on with the advent of TV, some families, and increasingly all families, had one TV set. It was in the living room, the centrepiece of family entertainment and information. There was one per household. As programming diversified, the issue of children wanting to watch something different from what the parents were watching started to introduce the second set to the home. Gradually families approached the ratio of a TV set per person.

Nowadays, single people in advanced countries can very easily have two or three TV sets in their home. One primary set in the living room, another TV in another popular room such as the kitchen or bedroom. We find nothing strange with the concept of one person having two or three TV sets. It is even possible to 'follow' two programs at the same time, as some fanatic sports enthusiasts might do at a time of intense sport play-offs, for example when two exciting games might be on at the same time, and the viewer would bring two TV sets into one room — or have one of those sets that allows the viewing of a small TV screen inside the bigger one on the main screen.

VCRs and other TV program recording devices have allowed us to shift the time of viewing programs, further allowing a person to consume content from two channels. The relevant statistic is how much content is consumed rather than the number of TV sets. Today in Western countries it is increasingly reported how much time is spent in front of the TV set.

Service Adoption

Figure 11.2 Subscriber penetration and growth rates, 2003. Even with high penetration rates, annual growth is still strong (Informa Global Subscriber Database October 2003)

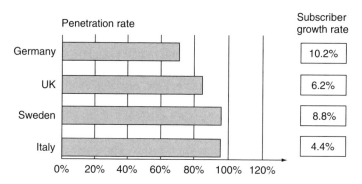

11.2.2 But can you use two phones at the same time?

The advent of multiple subscriptions to mobile phones introduces a similar pattern. Many young adults easily have two, some three, mobile phones. They can and do use them sometimes simultaneously — for example, speaking on a business conference call on the business call, and sending a personal SMS on another phone. The example shows that speaking on one phone, while sending and receiving SMS traffic on the other is a perfectly functioning means of communicating and is increasingly common in the advanced mobile telecoms countries. The third phone could even be receiving a voice mail message all at the same time.

11.2.3 Subscriptions and subscribers

Mobile operators report subscriptions and, until the advent of multiple subscriptions with the same person, a subscription could very easily be equated with a 'subscriber' or person with a mobile phone. Now with multiple subscriptions, it means that when all subscriptions are added together, the number is greater than the total number of people who own cellular phones, as some have more than one. The problem is difficult to measure very precisely as these multiple subscriptions are almost invariably on different networks, and the subscriber information is a closely guarded secret with mobile operators.

Machine subscriptions make the whole issue almost moot. As increasingly monitoring devices, remote control systems, remote payment systems, vending and ticketing machines, as well as all kinds of machinery from photocopiers to streetlights get embedded with cellular phone circuits and the SIM (Subscriber Identity Module) cards, the whole idea of a machine subscriber makes counts of subscribers all but meaningless. Machine populations of SIM cards are expected to exceed the human population well before the end of this decade. An alarm system that sends a hundred bytes of information per year is hardly comparable to the person who talks 5 minutes every day and sends three SMS text messages.

From the point of view of selling 3G services (or phones), it makes little difference if a person has one or two or three phones, and the subscription amount becomes less and less the key statistic. How many people are subscribers whether prepaid or post-paid, whether having one or multiple phones, is rather insignificant. The type of subscription will determine how that handset is used. It is the subscription that gets billed (or has its prepaid account deducted for billing), and the subscription is that which consumes the services and content. Therefore, if we look at saturation levels, we need to look at them from subscriptions, not subscriber numbers or handset sales.

The real revenues and profits are made with traffic. The amount and type of usage will increasingly become important. Telia, the Swedish incumbent when Sweden achieved 85% mobile phone penetration, when faced with analysts asking whether saturation was imminent, replied that on the contrary: only 15% of all voice traffic is on the mobile network, therefore the saturation is only at about 15% and has massive amounts of growth potential.

11.2.4 So where is the ceiling?

Recently many have speculated on the supposed 'saturation level' of Western European mobile telecoms. Yes it is true that many Western European countries already have subscription penetrations of around 75–80% and leading countries have rates beyond 90%. Conventional wisdom — and that relying on an understanding of fixed telecoms only — suggests that very soon saturation 'has to happen'. Penetration of 75% means that the 8-year-olds have phones, and there is a limit to how low that number can go. We cannot have toddlers age 3 with mobile phones, can we? If not, then we cannot have much growth left in penetrations. Also since it was true that we could survive with one fixed telecoms line, is easy to make the hasty generalization that we would be satisfied with only one mobile phone subscription.

11.2.5 'Near saturation' myth

The 75% being 'near saturation' is a remarkably widely held misconception even with seasoned telecoms experts. In fact, that 'saturation level ceiling' number has been creeping up as the numbers have forced re-examination. In 1998 the ceiling was widely quoted at 55% to 60% until Finland crashed through the 60% penetration level. Then the limit was argued to be 65% to 70%, and now many still hold the ceiling at 80%. The thinking was that children age 9 and grandmoms age 70 won't need or use cellular phones, much less advanced services like SMS text messages.

By early 2003, Austria, Finland, Iceland, Israel, Italy, Hong Kong and Sweden have seen their mobile phone penetrations go to the mid-90% level, with Taiwan having gone past 105% — and *each* of these countries has subscription numbers *still growing strongly*! As we saw earlier, as users are starting to have two mobile phone subscriptions, there is no inherent reason why the limit should be anywhere near 100%.

11.2.6 An American consideration

Those of our readers who are based in America or some of the countries where the mobile telecoms billing is 'Receiving Party Pays', i.e. that for all calls that arrive on a mobile phone, that phone's owner needs to pay for the call, those readers may be suspicious of the concept of two phones. It is very common in America for users to keep their mobile phones turned off completely, and for the users to carry beepers or pagers so that if someone wants to call, they first send a beeper message asking the person to turn on the phone (or to call). If a user carries one mobile phone but keeps it turned off most of the time, there is no reason of course to get another cellular device to carry along also, keeping it turned off as well. For those readers it is important to know that in most of Europe and many parts of Asia, and increasingly in Latin America, the billing system is the other way, with 'Calling Party Pays'. In that model, if you own a mobile phone, you are not penalized for receiving calls. That allows users to keep their mobile phone on always, no need to carry beepers, and all at once, it becomes feasible to have multiple phones. 'Calling Party Pays' is perhaps the biggest key to building the addiction to the mobile phone and the sooner the remaining

'Receiving Party Pay' countries like the USA abandon that outdated payment model, the sooner the whole industry will benefit.

11.2.7 How high is high?

The mobile phone penetrations are not about to level off any time soon, definitely not anywhere near 75% or 80%. Human subscription penetrations in Western countries will soon reach 120%–130% (almost all age 8–80 have one, half of 20–40 have two, and some have three). When you add machine penetrations like car telematics and home remote control and alarm systems, the machine penetration will be double that of human cellular penetrations. Beyond humans and machines are still pets and farm animals, and already some mobile telecoms solutions are being developed to address these markets.

We want to stress that it is prudent to reconsider the growth prospects of the telecoms industry, especially in light of the heavy 3G investments. Still, the discussion and analysis must be based on the *facts*, not on outdated misconceptions. The facts state categorically that 80% and 90% penetrations in Western European and Asian countries are a reality, and each of those countries grew at a healthy rate last year, and is expected to grow still this year. The expert consensus view in those advanced countries is that their penetrations will exceed 100% within a few years. The only question is how much over 100% will we go.

11.3 Business or Residential

The 3G industry has spent a considerable amount of time and energy on debating whether business customers or residential customers will drive the early usage of 3G. The very short answer obviously is that it depends on the mobile operator and will vary by country, so there is definitely no absolute correct answer. Not to mention that as we saw in Chapter 3, modern mobile telecoms customers are often both business and residential customers while using the same phone and subscription. We have multiple mobile phone personalities and can behave very differently during office hours and on weekends, etc.

11.3.1 The case for business customers

Business customers tend to be early adopters of high technology. Almost any technology will initially be targetted at, afforded by, and used by, business

Figure 11.3 Mobile purchases provide opportunities for up to 42% of all B2B transactions when counting travel and spot purchases

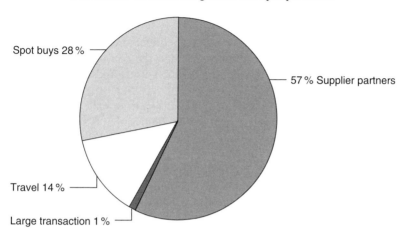

Source: Accenture Future of Wireless, 2001

customers. Microwave ovens first went to restaurants and hotels; video recorders to TV production houses and security companies; personal computers to investment banks and accounting firms; cameras to journalists, artist and the military, etc., etc., before any started to enter the residential home.

Businesses can make serious business benefit calculations on new technologies and their expected gains to the company's efficiency, competitiveness or service offering. If the payback period is reasonable for the industry and the company has access to financing, there is no real limit to the technological investment, and purchases such as 100 million dollar jet airliners can be bought by the dozen by major airlines.

Closer to the 3G area, with telecoms in general and mobile telecoms in particular, the mobile phone was first adopted by businesses. The early phones were cumbersome, expensive to buy, very expensive to use, and no real mass-market applications existed other than voice. Also the price differential between fixed-line voice and mobile voice was nearly prohibitive. The mobile phone was quickly adopted by some business people with an exceptional need to be continuously reachable, or who were exceptionally mobile. Investment bankers, top lawyers and management consultants started to carry cellular phones for urgent contact needs. Travelling sales representatives soon noticed that they could conduct much business from their cars and on trips to and from airports. Then the gadget started to achieve executive toy status and became a 'must-have' accessory for the yuppies.

As 3G will mean replacing the current phone, and as 3G is likely to be more expensive for at least a few years than the equivalent 2.5G phones, then the strategy of approaching the business users and offering them 3G phones is an attractive one.

Many of the early 3G service propositions seem to have early attraction specifically to business. A group calendaring solution that is always up-to-date in real time, can have some utility within the family, but the same solution will be very useful in any corporate team, group, department, etc., setting. Video calls are likely to face initial doubt and rejection by the mass market, whereas video conferencing is already becoming familiar in the fixed telecoms side of the business world. Few homes have family intranets but most businesses of almost any size tend to have them now, and secure and fast wireless access is a valuable use. We could go on and on. Perhaps the biggest single argument for 3G to start with business is that it avoids the chicken-and-egg question. With mass market residential customers there is no real utility to having the first 3G phone until enough of the population has them. There is little reason to develop content and applications to 3G phones if the users do not yet have the phones. Until there are users, there will be no network utility and no attraction to generating mass market content and applications for 3G.

Business customers do not need to worry about mass market adoption of 3G phones. If a given business finds a benefit then it only requires the financial decision to acquire the technology, and of course then integration with existing IT systems.

11.3.2 The case for the residential customer

While the telecoms industry did see most technical innovations being first adopted by businesses, such as answering machines, fax machines, speaker phones, etc., significant recent mobile telecoms innovations have addressed the residential customer first. SMS text messaging was adopted first by the younger generation who were not of working age. They then taught their older brothers and sisters, their parents, and later their bosses with their first jobs to appreciate how efficient SMS text messaging was in its speed, secrecy and situational convenience.

Service Adoption 177

Most business customers are not eager to deploy another new IT/telcoms solution. With numerous concurrent and conflicting needs already pushing into the business world (such as the PDAs, Wi-Fi (802.11 or W-LAN) modems for laptops and wireless networks for offices, corporate centralized SMS management for multicasting and desktop to SMS traffic, etc.), each new technical solution creates possible incompatibility problems with the ever wider range of legacy systems. Does the new solution work with the current e-mail system, how about the firewall, is it secure, etc. Will it require new software purchases? Are these compatible software or the same software as is already in use. Even if the same software, will it be a different release that will again cause the same headache across all of the existing legacy systems. And so forth.

At home we do not have multiple legacy IT and telecoms systems that are somehow integrated or interconnected. Yes, we have our stereo set and TV-video system, but for most residential customers today there is no digital link between these and, for example, the old PC for the children's homework, web surfing and gaming use. So bringing in a new 3G device will not cause disruption to the existing interconnected systems of the home.

The mobile operator does not know business customers very well, but knows the residential subscriber generally much better. The mobile operator has better sales channels and service support for the residential customer's needs than any for business customer's needs. The mass market holds large potential for fast returns through tools of mass marketing, such as TV ad campaigns and store-promotions. These are familiar to the mobile operator, whereas each business customer deal would require its own separate sales effort. And with new technology, quite possibly the business customer would also insist on some utility calculations specific to its type of business. The sales would be slow, whereas if the marketing campaign worked well, a new generation of phones could be very rapidly sold into the existing customer base. For example, J-Phone in Japan converted over half of its 12 million customers to new camera phones in a little over a year. It can be done, goes the argument.

11.3.3 Exceptional issues with 3G

The choice becomes harder with 3G than it was with, for example, the introduction of WAP and GPRS. Most concepts for early 3G handsets seem to be very feature-rich. With built-in cameras, colour screens, built-in utility applications, large amounts of memory, etc., 3G phones will replicate or even exceed the feature sets of the most expensive existing phones. Therefore, without considering the added costs of the 3G technology and that there has been no effect of efficencies of scale, a 3G phone will be among the most expensive handsets in the marketplace. This speaks against residential customers. Of course, handsets could be manufactured with small monochrome screens and minimal features but that would strongly defeat the purpose of a 3G phone to begin with.

The 3G network will not be perfect at launch. There will be both problems of radio network coverage — where the parallel 2G GSM or CDMA network will work just fine, the 3G network will experience dropped calls, etc. — and with the handover between 3G and 2G there are going to be problems that only 3G phones can experience (a 2G phone will not have any need to try to switch to the 3G network as invariably the established 2G network will have better or as good coverage as the new 3G network.)

The dilemma cannot be resolved in this book. If the mobile operator is strong in the business customer market and has learned about its key customer groupings through the upgrades to WAP and GPRS, there is a good foundation for then offering 3G to those who can be found to be most responsive to the new benefits. If the operator is strong with the more innovative and experimental segments of the consumer market and has educated that market to adopt messaging, mobile commerce, etc., on SMS, WAP and GPRS, and in particular if the segmentation system can identify likely prospects,

Figure 11.4 Communication technologies by user group plotted against adoption cycle (S-curve)

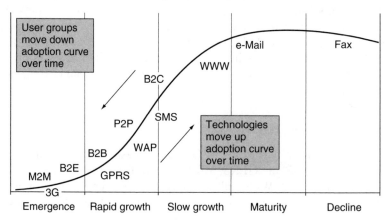

then that is the obvious first focus. For a newcomer or 'greenfield' operator it is, of course, most difficult where little, if any, information exists on the true users and usages of 2.5G, and new and unproven technologies need to be marketed and sold. Regardless of where the initial focus will be, the actual first converts to 3G will be early adopters by their very nature, whether residential customers getting 3G for themselves, or business-customer decision makers with a willingness to push the technical envelope. So we will need to examine early adopters more closely.

11.4 Early adopters

One could say that, sequentially, any person who adopts a new technology among the first users, is an early adopter. In a strict chronological sense that may be true, but early adopters are also a user type that needs to be understood. The early adopter is not a typical user, and actually the early adopter is remarkably different from a mass market user. The early adopter is interested in technology, often has a multitude of gadgets and usually is very well aware of the latest technologies. The 'techno-geek' and 'gadget-freak' early adopter can be a telecoms-services sales representative's or calling centre person's worst nightmare, by asking about the most obscure facts, fully aware of the current arguments on compatibility issues, etc. However, the early adopter is absolutely crucial to the overall success of any new technology. In his report *3G Launch Strategies: Early Adopters Why and How to Make them Yours* (2002), Steve Jones proves that the reception of any new technology by the early adopters results in the eventual total success of the technology in question. Both the ultimate penetration of a new technology, and the speed of its adoption are determined by how rapidly early adopters embrace the technology (Table 11.1).

The early adopters are passionate; they usually identify a need for which they want an emerging technology. They are quite willing to accept imperfections in systems and solutions for the overall benefit of having the latest. Early adopters are not attracted to mass-market publicity campaigns with sports heroes and glamour movie stars promoting phones, so eliminating the need to waste early marketing money on mass-market campaigns. They are near pointless anyway when early markets will see shortages of mobile phones and considerable price differentials against 3G.

Early adopters are not attracted to mainstream, mass-market applications like music, sports, games, gambling, adult entertainment, etc. Not that early

Table 11.1 Number of 3G early adopters

Country	Mobile subscribers	3G early adopters
UK	50 M	7 M
Germany	60 M	9 M
France	40 M	6 M
Spain	30 M	4 M
Sweden	7 M	1 M
United States	130 M	20 M

Source: Tariffica, *3G Launch Strategies*, August 2002

adopters would not use those services as well, but being very sophisticated technologically, early adopters tend to have specific technology needs they want to address with the new gadget. These may be converting some files, a specific condition of accessing e-mail, etc. Not the mainstream mass market uses. Best of all for 3G, the early adopters are evangelists for new technology that they have found useful. For the early stages of a new service launch, the opinions of the early adopters are vital, and if they decide to use their networks and spread the word, a service can move very rapidly from the early-adoption phase to the early majority phase when the service becomes mass market.

Attracting early adopters to any new service or product is a specialized marketing challenge. It differs radically from the mass-market approach for mass-market services. The early adopters expect to be treated especially well, and they expect higher levels of understanding from their sales outlets than do mass-market customers. The early adopters are willing to sustain imperfections in new services but expect their issues to be heard and the faults to be fixed in due time. Where mobile operators have been slow to adopt general marketing tools and methods, their track record with early adopters is simply horrible. Here is where mobile operators can use a large amount of support and assistance by partners, experts and consultants who specialize in early adopters. The success with these pioneers and evangelists is crucial in determining how well the company's services will feature in the overall market to follow. In his study Jones (2002) states that early adopters are important because they: 'provide higher profit margins; speed up the sales cycle; attract 3rd party

Figure 11.5 Delay from imperfect new service launch targeting

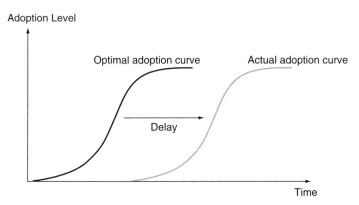

investment; generate positive publicity; and can be easily recruited at relatively low cost'.

11.5 Mass market

The mass market customer, the early majority, the late majority and the laggards, are very different in their attitudes to new technology. They are not willing to read through inconsistent user manuals and load updates from the web. They demand that help desk support at calling centres will know exactly what the service is and how to fix it. Mass market users will demand that any new service works the first time it is taken 'out of the box' and that it functions correctly 'always'. The mass-market customer will need very different levels of service reliability and ease of use.

The mass-market customer will convert any service from one serving a niche market to that of a mass market. The overall economics of service usage come into play. Also of significant issue is what is called Metcalfe's law (Robert Metcalfe is one of the founders of 3Com). This law states that the utility of a network will increase by the square of the members connected to the network. Thus any cellular network based services are likely to experience considerable added utility when numerous users start to use the service. This has been seen for example in e-mail and SMS messaging, and the same effect is likely to be soon witnessed in picture messaging as more mobile phone users acquire picture-capable phone handsets.

Figure 11.6 Suitability of digital services to mobile phones

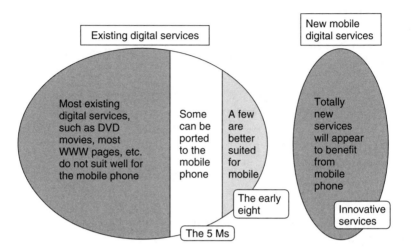

11.6 The early eight

Numerous services have been introduced on mobile services on SMS, WAP, I-Mode and GPRS around the world. While some services have surprised early analysts and exceeded the wildest dreams of the industry, many more seem to have fallen flat, with service adoption lagging behind the early predictions. The mobile phone is very different from other digital delivery media such as the CD-ROM, the fixed Internet, digital TV, etc. The mobile phone is a poor competitor to surfing the Internet when compared to broadband Internet, and is equally a vastly inferior tool for consuming full-length motion pictures. However, as is typical with any new tools, there are strengths as well as weaknesses. Now, for the first time, eight classes of service have emerged where the cellular phone has an absolute competitive advantage over all other service delivery options. These 'early eight' are obvious candidates for new service development for any cellular operators.

The first of the early eight service types is *micropayment* purchases, meaning any purchases worth less than 1 dollar/euro/pound. Whether buying goods from vending machines, paying for stamps, bus tickets, parking, tolls, any kind of small payment, no other device or payment mechanism provides such convenience as the cellular phone. Coins are the nearest rival, but coins require handling and any vending machine or other automated system will need constant emptying of coins. Coins pose a problem in that they are not always readily handy in right amounts. Credit cards and banking debit cards are cumbersome and credit cards especially require minimum payments. Credit cards are not available to the youth. The billing system of the mobile operator can easily handle micropayments and countless micropayment purchase solutions already exist. The competitive advantage for the user comes out of the fact that even if one has run out of cash, one always has the mobile phone. And regardless of country and local currency, the mobile phone can handle the transaction.

The second early-eight service is *any digital content that is consumed on a portable device*. The most obvious services are games, music and digital photo images. Very many games already exist on versions that run on handheld devices, but often these are hampered by limited resolution on the screen and small amount of memory capacity in the gaming device. With today's and tomorrow's technical capabilities of mobile phones, an excellent gaming platform is in every pocket. The same is true of portable music players. Walkmans, portable CD players and minidisc players require size and weight to handle the storage medium. MP3 players on the other hand usually need a separate connection to the music source. The natural environment for digital music is to download it to the mobile phone for consumption on it. Also, separate from the fixed Internet, the mobile Internet provides a much more secure delivery system to track intellectual property rights. On digital cameras there will probably always be better cameras technically that are 'only' cameras, than those that are a compromise combining the mobile phone and camera. Increasingly for mass markets, the primary camera will be that on the mobile phone.

Table 11.2 Worldwide gaming software revenues, 2003

Console games	16 500 M USD
PC games	3 800 M USD
Handheld games	2 200 M USD
Mobile phone games	590 M USD
Interative TV games	250 M USD
TOTAL GAMING SOFTWARE	23 000 M USD

Source: *Wall Street Journal*, 17 October 2003

Third on the list of the early eight is eliminating instances of people *standing in lines*. Most occasions where people stand in lines to buy concert tickets, movie tickets, bus and train tickets, airline tickets, sporting event tickets, etc., can be handled on the mobile phone. While the fixed Internet arguably could do that as well, in most fixed Internet ticketing solutions the person still has to pick up a ticket at a designated counter against a confirmation number or the like. With mobile phones it is possible to have the electronic ticket delivered onto the mobile phone and validated on the phone, removing the need to stand in a line to pick up the ticket.

Next on the early eight, is the sale of *intangible services*. Any case of truly intangible services can be, and should be, sold on the mobile phone. Any service where no goods are ever delivered are perfect candidates for this. It includes lotteries — the lottery ticket is only proof of partiticipation, but the selection of numbers is an intangible service — no numbers are actually given to the buyer. In fact all gambling and betting is intangible. Insurance, especially short term insurance such as travel insurance, is perfect for sales on mobile phone. Licences, a fishing or hunting licence; taxes and fees for the government; school fees; anything with intangible services, can be sold on the mobile phone, and the buyer can have the proof of purchase delivered to the mobile phone.

The fifth of the early eight are *personal services*. The more sensitive and private the personal service can be, the more it suits the mobile phone. At one extreme is adult entertainment, which will mimic the local legislation and norms for decency from the mild such as swimsuit calendars to the hard core that is considered acceptable in Northern European countries. Services that address personal issues can be much more, however, and can include such areas as dating, medical services, credit and finance related services, etc. Personal e-mail, messages and chat with teenager siblings is better suited to a mobile phone than the family PC, as the teenager need not fear that the little brother breaks into the e-mail and reads what big sister has been writing about.

Next on the early-eight listing are *location based services*. In those cases where the very core of the service proposition is dependent on location information, the mobile phone is unbeaten. Typical examples are tourist guides and location-specific maps and any kind of 'Find Me' service, such as find me the nearest cash machine, the nearest hotel, the nearest petrol station that is open at this hour, etc. Location based services can also include finding your friends in a busy place, locating the kids, even tracking your car or your pets.

Figure 11.7 Location Based Services will form a key addition to service revenues in several service types by 2005

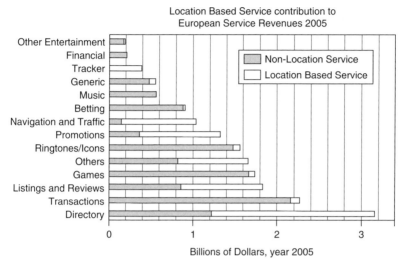

Source: The Yankee Group, 2001

The seventh of the early eight is human *interaction with machines*, especially if speech is involved. These types of service range from listening to messages — the voice mail is an early example of this — to voice prompts, voice recognition and speech synthesis. One of the most advanced voice recognition and speech synthesis solution has been introduced in Japan by the KDDI network, where automated translation is now possible between English, Japanese and Korean. The person calls the service, speaks in one language, the speech recognition understands up to seven seconds, performs the translation, and then uses speech synthesis to speak in the desired language. These types of solutions will soon expand rapidly. The mobile phone is also the perfect 'remote control' for any device or need that is beyond the immediate vicinity, such as turning your home air conditioning or heating on remotely, checking on a remote security camera, or controlling automation and robotics.

Last but not least of the early eight is *urgent information*. Whenever time is truly the driving factor, the mobile phone provides the best tool for delivering the news. Whether it is a stock quote update, a sports score or a breaking news story, the fact that the mobile phone is always within hand's reach makes it the fastest media. Numerous user-developed solutions have proved the incredible speed advantage as a message via mobile phone can reach people in meetings, far from their e-mail, even when radio coverage is so patchy that voice calls cannot be sustained. The mobile phone messages go through. Busy executives already use SMS text messaging as a means to make last-minute changes to daily schedules and meetings. Their secretaries and colleagues have learned that if the busy exec is not answering the mobile hpone, has not had time to read a fax, listen to voice mail or read an e-mail, the fastest way to alert them is again, via an SMS message.

The early eight is not a collection of killer applications for the general public. Any one of the services described might become a killer application for an individual user, and a combination of them might form a killer cocktail for even significant segments of the population. However, the early eight are not necessarily going to be the biggest revenue sources nor produce the heaviest traffic. What the early eight do provide is a grouping of services where the mobile phone holds an absolute competitive advantage over any other delivery means or media channel such as the CD-ROM, digital TV, broadband Internet, etc. It means that for the end-user, when the service has migrated to the mobile phone, that user will observe an overwhelming benefit over the previous ways of accessing the service. The early eight are therefore services that are inherently natural for consumption via the mobile phone, and the change in behaviour in our society is easiest to generate using the early eight. A prudent cellular carrier/mobile operator/ wireless service provider will deploy early eight services the soonest, to illustrate to its customers that the mobile phone can deliver superior service.

11.7 Beyond the adoption

This chapter looked at how technology is adopted by developing the S-curve and its thinking. The chapter illustrated how even at 80% penetration levels and beyond, our industry is still not at saturation. Then we discussed business and residential customers and then showed how early adopters differ from the mainstream and mass market customers. Finally we offered the theory of the early eight to identify classes of services that will be rapidly adopted by mobile phone users. The road to make money with new technologies, no matter how intuitively excellent they may be, is always bumpy and unpredictable. Who will and who will not be able to make money eventually with 3G is still open to fierce arguments. An insightful thought was published by the *Economist* when looking at another dramatic technological development — the introduction of electric light — back in 1881: 'The electric light is very probably a great invention and let us take it for granted that its future development will be vast. But this, unhappily, cannot be urged as a reason why the pioneer companies should be prosperous'.

There are three kinds of death in this world. There's heart death, there's brain death, and there's being off the network.
— Guy Almes

12

Reachability
Why Mobile is so Different from Fixed

Early on, when the mobile phone first appeared, it was thought of as a wireless telephone. Its use was considered in the same way as the fixed phone, and a straight substitution of the same services was assumed. Before the mobile phone it was not possible to imagine how drastically this device would alter behaviour and create new phenomena. The concept of reachability is the most relevant of the new business opportunities presented by mobile phones. Reachability is the key motivator for the change in customer

---

3G Marketing: Communities and Strategic Partnerships Tomi T. Ahonen, Timo Kasper and Sara Melkko
© 2004 John Wiley & Sons, Ltd   ISBN: 0-470-85100-7

behaviour that is being witnessed worldwide. However, before we look at 'reachability', let us see how two other significant technological developments were first received: the car and the fixed-line telephone.

12.1 Wireless carriages and voice telegraphs

By thinking of the cellular phone as 'only a wireless substitution of the fixed phone' the industry looked at mobile phones much like the horse and buggy industry looked at cars — as 'horseless carriages' — and as the telegraph industry looked at the first fixed-line telephone: as a 'voice telegraph'. While it is true that a car can replace a horse and buggy, and that voice communication on the telephone could transmit messages replacing telegraphs, both substitution assumptions missed out on the more important aspects of the inventions.

With cars, man had a means to travel for leisure. Before, in the times of the horse and buggy, it was not possible to 'ride' to another city for a vacation, unless one was a true adventurer, or willing to ride horseback for days. With the car it became possible to drive from New York to Florida, or from Paris to Nice and enjoy a car holiday. Certainly the car also replaced the common rides about the town as were done with the horse and buggy, but in the late 1800s, nobody could guess that large parts of the population would travel to beach resorts in other parts of the country. Whole new industries evolved around the car that were not promoted by the horse and buggy. Some were necessary to the new technology, such as petrol stations and automobile repair garages; others served the new behaviour made possible by cars, such as motels (motor hotels), drive-through fast food service, etc. It was not until the mindset of replacing the horse-drawn buggy was removed, that new industries could be invented around the car.

With the telegraph, mankind had a means of communicating across vast distances, but the return communication was not immediate. Telegrams were 'one-directional' contacts: I informed you about something, then I waited to see if I received a return communication, perhaps today or tomorrow, from you. Again, with the 'voice telegraph' man could use the telephone to make statement communications: 'I will be arriving on September 4 at Singapore', but again, nobody could expect that people would start to 'talk' on the phone, that people could call each other up only to chat and find out how the other was doing. This was not the mode with the telegraph, and of course today most telephone conversation, at least in terms of time used, is interactive sharing, not strictly of an informing nature. It is very common for close family and friends to call up each other 'just to hear your voice'. Where the initial telegraph was an expensive communcation tool, the telephone soon became a social instrument, but this change also needed people to move beyond the idea that telephones merely replaced telegraphs.

12.2 Enter reachability

Nokia end-user studies have calculated that it takes from 4 to 6 years to change behaviour. Ericsson has measured it as taking from 5 to 7 years, and a similar finding comes from Visa, who have observed that it takes from 3 to 5 years from the time a person first goes onto the fixed Internet, and the time the user is comfortable placing purchases on a credit card on the net.

It is not until we see the changes in behaviour from a new technology that we can start to evaluate that technology's impact on society as a whole. The first changes in society related to the mobile phone are of course visible in the early adopting countries where the mobile phone first became a mass market product during the mid 1990s — mainly Finland, Sweden, Denmark, Norway and Italy. One of the most significant societal effects with a direct impact on the business of mobile telecoms is called 'reachability', possibly first identified in Finland as 'tavoitettavuus'.

Reachability started to enter the mobile telecoms vocabulary among marketers in the late 1990s in the early adopter countries. Reachability allows the cellular phone user to be contacted, reached, whenever the user wants. This was not true of the fixed telephone. With the fixed telephone you called someone's home or office, and if the person was not available, left a message. The message might be left on an answering machine, or with a family member if called at home, or a secretary or work colleague if calling at work. With fixed-line phones it is quite common for another person to answer the phone — such as a family member at home or the company receptionist, operator or secretary at the office number. Older generations from before mobile phones have been conditioned to answer any ringing phone and these generations will also find it annoying if someone chooses not to answer the mobile phone. The fixed line phone is not strictly personal, and as often as not, you are not near your fixed phone(s).

With mobile phones the device became personal. We carry it on our person all the time. As penetration reached 80% and above, all who knew how to use a mobile phone also had one. As it is personal and carried on our person, the mobile phone no is longer answered by others. Very importantly we have learned to keep our phone turned on whenever we are willing to

Figure 12.1 Evolution of the mobile service user

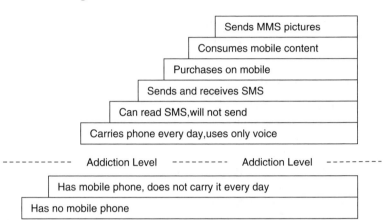

accept calls and messages, and turning it off if we want privacy from mobile phone intrusion. People of the ages of 30–50 have a hard time reconciling emotions about not answering a ringing mobile phone. Younger generations have no such qualms; they happily leave incoming calls unanswered, returning calls selectively when it suits them.

12.2.1 Calling the person, not the place

The first aspect of reachability is that with mobile phones we call a person, while with fixed phones we call a place. When we call a fixed wireline phone it is typically a home or an office, and we are accustomed to hear someone else answer the phone and ask us who we want to talk to. At home it may be the mother of the family or a teenager. At the office it may be the switchboard or secretary rather than the intended person. With the mobile phone it is hardly ever someone other than who we intend to reach. We are very much surprised if a familiar mobile phone is answered by anybody else than its normal user.

Reachability changes the nature of the contact and also changes the type of conversation. On the mobile phone we have more intimate conversations. We can walk with the mobile phone to another room, into a spare office at work, step outside at home to achieve greater privacy for our conversation than is possible on a fixed-line phone. As we reach our intended person through the mobile phone, it also means that conversations on mobile phones are more personal and more immediate — we know who will *always* answer that phone. In cases where we want to be cautious, discrete even secretive, we do not want to call a *place* where anybody might answer a fixed wireline phone, we call the mobile phone of the person. With calling line display, in most cases the recepient also sees who is calling, so there is less need to identify both in calling. Again many people have stopped identifying themselves when answering the mobile phone, and greeting the caller by name, such as answering the phone with 'Hi Joe, what's up?'

12.2.2 Change plans

Reachability allows changes of plans. As mobile phone density increases, ever more people will have mobile phones. With the inevitable last-minute changes in our everyday plans, we soon learn to inform the other people of our being delayed. We use our mobile phone to place the call, as we are rushing and cannot stop to use a fixed wireline phone. Moreover, the people we want to inform of our delay will have their mobile phones, and we call them. With early adoption of mobile phones it will be someone in a group who has a mobile phone, and we call that person. As penetrations increase, ever more of the total population have mobile phones and we expect to be able to call anybody to inform them of changes in plans.

As mobile phone users intuitively discover what we call Reachability, they expect everybody always to be connected and ready to receive last minute warnings of changes. Forgetting the mobile phone is no longer an option. So if your friend is running late, the friend will call you on your mobile phone to tell you. As we learn to make updates via mobile phone, we practically insist that our friends also carry their mobile phones at all times. Because it enables both parties to make and receive updates, the mobile phone is seen as indispensable as a scheduling tool. Peer pressure and changes in culture force all to conform.

Reachability then brings about changes in our non-telephone behaviour. As we learn that most of our friends have mobile phones anyway, and that they can be reached at practically all times, we can develop further flexibility in our *keeping of* schedules. Now the mobile phone becomes a tool for allowing last-minute changes. We can send that other e-mail because we can then call our friend to say that we are 5 minutes late. We can stop by the grocery store on our way home as we can call the kids to let them stay 10 minutes longer at the park. The mobile phone, and reachability, has

Figure 12.2 Growth in SMS use in the UK 2000–2003 (Mobile Data Association September 2003)

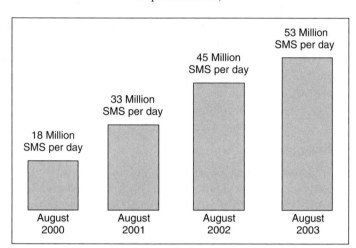

altered our sense of what schedules are fixed and when. Reachability gives us flexibility in our daily chores.

12.2.3 Indispensible

Reachabilty makes the mobile phone ever more indispensable. Now we need it in case we have a change, and very importantly, reachability means that someone with an unforeseen reason can contact us and deliver the important news immediately. As we discover how reachability is valuable to us both in that we can reach others and others can reach us, it makes us more dependent on — one could say addicted to — the mobile phone. Reachability also makes us proponents for promoting the continuous connectedness of all of our friends and colleagues. We can never guess when the sudden need may arise to contact another person urgently, but when that happens, we want the other person to have the mobile phone with them and turned on, even if in silent mode.

12.3 Reachability and mobile services

Reachability is one of the distinguishing aspects of mobile telecommunications. It does not exist in fixed telecoms nor on the fixed Internet. Much like those who noticed the opportunity from the new behaviour in horseless carriages, and started to build motels and service stations, and those who noticed that voice telegraphs allowed creating new businesses like Yellow Pages advertising, those who understand reachability can introduce services to build upon this new need.

Some of the early and obvious services that address reachability type issues stem from the fixed telephone times, when reachability was a subconscious need. The answering machine and its various network and service variations of voice mail, etc., are early business ideas serving an unmet reachability need. One of the drivers for rapid SMS text messaging growth has been reachability. A survey of British executives by the Mobile Data Association in 2002 found that their most common use of SMS text messaging was about meetings and schedules.

12.3.1 SMS text messages and reachability

We often hear the expression that time is money. Now with the funny little service, SMS text messaging, the industry discovered the mobile killer application as this communciation method was the fastest possible means of communication — often bettering voice calls and their voice mail ping-pong by hours. SMS was the first new service fully to serve our previously unspecified and subconscious need for reachability

Although SMS implies restrictions for the users — such as limited message length and relatively high prices in some market areas — it has proved to be a big success. The constraints of a service can be both vice and virtue. Now with dramatic growth in its usage, a lot of research has been undertaken on SMS users: i.e. boys tend to write short notices, and girls are creative in finding ways to abbreviate their eloquent messages in order to shorten them and save unnecessary spaces. Then there are otherwise active mobile telephone users who just do not like to bother with small displays and buttons and therefore detest writing SMSs. Understanding the user's wishes and problems are the basis when creating solutions and new services.

So why is it that users are perfectly willing to pay high fees for a service like SMS, which has to be short and is rather complicated in use? At the same time the same users are reluctant to pay *anything* for sending e-mails that are able to carry much more information and can contain even audio-visual attachments. The decisive factor lies in reachability and time. The importance and relevance of this kind of short and instant messaging is that it is sent and received to a very high probability in nearly real time. The user wants to be sure that the very moment the urge to communicate occurs the message is conveyed to the addressee. The reachability factor is a stronger factor than any technically 'superior' messaging services that cannot match SMS speed. The other reason that SMS is read in the shortest possible time is — surprise — its shortness. With a limited amount of characters it is very hard to get spammed and there is hardly anyone that will think 'I will read that SMS later'. You always have time to read 160 characters at once.

On the fixed Internet e-mails tend to be associated with a fixed location. Often there are only one or two locations where most e-mails are read, such as one's home or office. With e-mail there are no guarantees that the message is received, much less read at short notice. In 3G, of course, e-mails too can be received in real time, which can make them as fast as SMS as far as reachability is concerned, if both sending and receiving party use a 3G device, and *both parties know* that the other is 'always connected'. With our e-mail overload and the fact that most e-mails are sent with poorly considered, incomplete and even misleading subjects, or simply the 'RE' for reply without any editing of the subject header, many important e-mails are ignored for too long. It is likely that SMS will continue to be the fastest form of communication even in the 3G world.

12.3.2 Respecting privacy

We also use SMS text messaging to meet a reachability need when not sure about invasions of privacy. This can be for example when a work colleague might not be in 'work mode', such as over the weekend or late in the evening. It can also be when a friend is suspected to be asleep or otherwise not likely to want to talk. In such instances it is usually safe to send an SMS.

Operators, service providers and application developers need to explore reachability, see how early-adopter users in high mobile penetration countries are using mobile services, and which of those services pose reachability problems. Any problem is an opportunity for a solution. Some early ideas of reachability services include finding people physically, locating their numbers or enabling filtered contact with those who want privacy, etc. Providing automated solutions to handle those cases where we want to be out of reach, but still want to be able to serve the person calling can also be quite useful. As reachability is a learned pattern of behaviour, that is developed automatically as a mobile phone user gets ever more experience of noticing the phenomenon, it is also important for mobile operators to start to educate all users in the benefits of being connected. The sooner they discover reachability the more they will consider the mobile phone indispensible.

12.3.3. Knowing who calls

Reachability also manifests itself in the way we can see on our mobile phone (in most cases) who is calling. Thus in most cases we can identify the caller immediately if we have have saved the number in the telephone's memory. It is up to the called person to answer the call at once or return it later. Since both the caller and the called person both know that the called person knew who called, even a deliberate decision to not answer is a form of communication.

There is of course a natural curiosity about unknown numbers. Services have been created to allow queries, for example, through telephone directory services via SMS to find out who is behind an unfamiliar number. This avoids the costly and slightly embarrassing 'um...sorry...somebody called my phone from your number...' type of conversation. Your mobile number is not a mere series of digits — it is your identification. Therefore the option to keep your own number when changing the operator is an important marketing tool for service providers. The number identification can even get so far that there are some people who refuse to answer 'unknown caller' calls as a matter of principle. A call is an intrusion into your privacy and you do not want to give that privilege to someone you do not know. As many say, 'I carry my mobile phone on my person to help me communicate, not so that you can cause intrusions in my life'. We make active decisions to include and exclude people to and from our personal communities through the mobile phone.

12.4 Cellular is a distorted case of Metcalfe's law

Reachability and our addiction to the mobile phone, and the ensuing communciation patterns, differ from the textbook theories of previous communciation networks. The growth patterns in traditional voice services on fixed telephone networks are at or near saturation — in some countries they are already diminishing, having peaked. Fixed telephone network traffic patterns are very predictable and tend to follow near-linear patterns. That is because the total number of connected voice callers has plateaued. With fixed telephone networks the utility and traffic patterns follow Metcalfe's law — the utility of the network increases as the square of the number of connected parties. (Robert Metcalfe founded the computer networking company 3Com.) Usually Metcalfe's law is expressed as an exponential curve as is shown in the following graph:

Figure 12.3 The standard Metcalfe's law produced a standard exponential curve, typical of most telecoms services

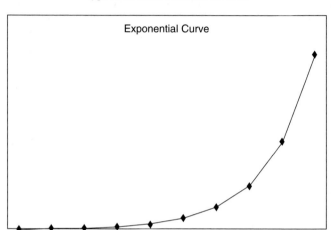

Metcalfe's law is based on the assumption that the value of the service is uniform to connected parties and Metcalfe's law has been well documented and accepted as explaining network traffic in fixed telecoms and fixed datacoms networks.

12.4.1 Hockey stick is not Metcalfe's law

The traffic in cellular networks initially seemes to follow Metcalfe's law until a certain 'critical mass' has been achieved, and then the cellular network traffic pattern breaks that which is predicted by Metcalfe's law. Cellular networks have a much more pronounuced change in traffic, producing a clear inflection point. The inflection point is dependent on subscriber amounts. The phenomenon was first noticed on GSM networks when the first countries started to experience high 20% and low 30% penetration rates. When this happened, those isolated incidents were dismissed as anomalies in the various early adopter countries of GSM, but when the pattern repeated itself across all cellular networks for voice, and started to repeat itself in the same markets for SMS text messaging, it was obvious that Metcalfe's law was not directly applicable to cellular networks.

Individual operators could not verify that this phenomenon occurred in other networks, so the first to notice the pattern were the network planning engineers at the equipment manufacturers. Cellular network engineers with their network equipment manufacturers mostly situated in countries like Sweden, Finland, Canada, Germany, etc., where ice hockey is also quite popular, very quickly named this phenomenon the 'hockey stick curve'. Some people from other networking technologies then proceeded to call all fast growing usage curves 'hockey stick curves'. In most fixed network cases the incidence is actually normal Metcalfe's law and to call such patterns 'hockey sticks' is not accurate. Most fixed networks conform to Metcalfe's law in its basic form, in an exponential benefit curve. Cellular networks do not. They may be a special case of Metcalfe's law but the shape is distinctly not exponential, it is more pronounced. Cellular networks are somehow a special case and further understanding was needed.

It is now clear that cellular networks introduced a new value element to the cellular network, which does not exist on fixed telecoms and fixed datacoms networks. As we have now seen, that new benefit is reachability. Reachability produces a distorted Metcalfe's effect, more accurately portrayed by the form of a hockey stick curve. Thus network utilization in cellular networks does not fit the traditional exponential curve as predicted by Metcalfe's law, but rather cellular network traffic and revenue have a more pronounced inflection point and the hockey stick effect is more dramatic.

Figure 12.4 Mobile services adoption takes the form of a more pronounced inflection point in what is called the 'hockey stick' curve

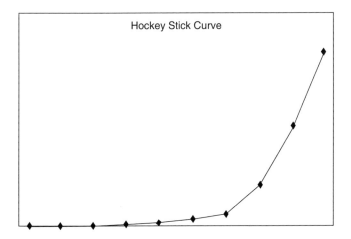

12.4.2 Inflection points for the hockey stick curve

The hockey stick effect has been observed and documented so far with cellular voice traffic, where the inflection point has been spotted at about 18% penetration rate. After that, the cellular traffic growth rate changes dramatically. This pattern depends, of course, on the reachability effect and thus requires that the phones are kept turned on at all times. Otherwise, as in the USA, the growth curve is much less drastic, following only the Metcalfe's law rate.

More recently, the hockey stick effect has also been observed and documented with SMS text messaging traffic. The first instances of SMS traffic growth reaching an inflection point was measured in Scandinavia in 1996–7. The early data was initially dismissed as anomalies until the same phenomenon repeated itself in most countries. Soon the inflection point for SMS was found to be at about 28% penetration rate. There were several countries that did not conform to this pattern, and the second requirement was set that there has to be SMS roaming between the operators/carriers within that country, for the 28% penetration rate to trigger the accelerated growth. With this caveat, all countries have displayed the same traffic pattern. More recent mobile services are too new to have produced enough data to verify a hockey stick effect and further study is needed to isolate their likely inflection points. An extreme example of how critical this point is, was seen in the USA, where in the summer of 2001 the total volume of SMS text messages sent was 30 000 per month. After each of the operators concluded SMS interconnect, by the summer of 2002 that number had shot to 1 billion SMS text messages sent per month. Unfortunately, the USA cellular industry had artificially depressed its SMS revenues for many years and essentially foregone billions of dollars of SMS revenues before 2002.

12.5 Most personal device

As we become increasingly addicted to our mobile phones, they become indispensible to us. We cannot imagine life without them. Mobile phones are the first thing to be turned on in the morning — or switched off when they act as an alarm clock. Mobile terminals are the only electronic devices that we carry around constantly. No other electronic equipment shares so much with you — it is there all the time and turned off or to silent mode only when you are sleeping. Some people keep the mobile phone turned on practically 24 hours, 7 days a week. This is why our expectations towards mobile phones are increasingly personal and even passionate.

Mobile phones are no longer only devices to call but they have multiple functions: address book, time scheduling tool and calendar, birthday reminder, play box, etc. How often do you have a look in your Filofax compared with your phone display? 3G terminals have to be able to communicate with other different software and machines like PDAs (personal digital assistants), PCs, etc.

The mobile phone is also increasingly an extension of our personality. The selection of our phone is an important decision — for most people, of course, but there are a few who loudly protest against this and their selection is equally a statement about how independent they are of their phones. We customize the appearance of the phone through interchangeable covers, and we purchase ringing tones to make our phone sound distinct and very much also to communicate something to our surroundings whenever we receive a call or message. We do not react so personally to our TV, toaster or CD player.

The function of the mobile phone in 3G is no longer mere calling: there are numerous new activities like the phone as a work instrument, an entertainment device, a locator for finding, a navigation tool, a verification method, a means to enable our purchases, a delivery mechanism for our mails and messages, a camera and viewer for digital images, a search engine, a sharing sharing tool, a music player, a gaming device, a life experiences depository and record keeper, and various other forms of action, reflection, or communication. Companies that understand how the mobile phone is now emerging as the supreme key and solution to our most basic human needs will be able to capitalize on these opportunities. At the very heart of our addiction to the mobile phone lies that magical phenomenon, reachablility, that the consumer can be reached and services can be offered, sold and delivered via these devices at practically any time of day.

12.6 Reach out and touch

Reachability is a new phenomenon in mobile communications. It did not exist in fixed telecoms and it did not exist on the fixed Internet. Reachability has already created opportunities for new services, and understanding reachability will bring many other new opportunities. For us to be able to capitalize on reachability we first need to understand it, but that understanding is worth it, as it gives us the ability to innovate in this area, as Louis Pasteur said: 'Chance favours the prepared mind'.

Any fool can paint a picture, but it takes a wise man to be able to sell it.
— Samuel Butler

13

Selling Mobile Services
Could I Interest You in Some 3G?

After the Attention, Interest, and Desire stages of AIDA have been achieved, it remains to complete the sales with Action. This chapter looks at how to sell 3G. Mobile operators and their partners have to develop functional sales strategies for selling the 3G services and products. In a very general sense this will be similar to selling 2G services and products, but as 3G services are likely to be of a dramatically higher level in variety, and in very many cases the services are more abstract or intangible, the whole sales concept becomes much more complicated.

3G Marketing: Communities and Strategic Partnerships Tomi T. Ahonen, Timo Kasper and Sara Melkko
© 2004 John Wiley & Sons, Ltd   ISBN: 0-470-85100-7

13.1 What do you sell in 3G?

Selling 3G can, initially, be a confusing idea. We need to be clear about what is being sold and to whom. To the end-user, 3G sales start off with a 3G terminal and subscription. Initially these two may be sold as a package, but soon subscriptions can be sold to people who already own 3G terminals (phones/handsets) and newer terminals can be sold to existing subscribers. Separately, to the end-user, 3G sales involves hundreds of 3G services.

The various players in 3G can also sell services relating to 3G to other parties, depending on the situation. For example, the network operator or mobile portal may want to sell 3G customer information data or 3G location information, etc., to various content partners and application developers. An IT integrator may offer 3G integration services to business customers such as mobilizing the CRM (customer relationship management) system, etc. In the areas of advertising and promotion there will be advertising services that can be sold, and with m-commerce, services such as credit/collection can be offered.

13.2 Selling through distributors

Initially, in mobile telecoms the services sold were very few and relatively easy for sales representatives in the distribution chain to learn. The early services included such items as voice mail, international calls, balance reminders, SMS text messaging, etc. With WAP and GPRS the range of variety in services exploded and increasingly the sales representatives were growing unfamiliar with all services. This diminished the sales representative's ability to convince the customer and close the sale. The various issues dealing with the distribution chain were dealt with in their own chapter (Chapter 7), so for us it is enough to keep in mind that with 3G and ever more advanced services, the matter of understanding what to sell becomes ever more critical.

When selling 3G services via distributors, packaging becomes crucial. The more abstract a product is, the more attention will be focused on the appearance of any supporting material that is physically tangible, such as brochures, any boxes or kits, etc. The packaging can vary from small sealed envelopes to book-sized product boxes to sit on the shelf at the store. The printing on the package is crucial to attracting initial interest in the service and should be designed also to serve as reminders for the sales representative who will need to sell the service. The envelope or box should contain detailed descriptions of the service and could add any of the following: a PIN-code, certificate, product flyer or brochure, any marketing gimmick such as a member-get-member coupon or coupons, and cross-sales material on other products that might be interesting to the target customer. Distributor support consumes time and resources and includes training, monitoring (is the service understood and is it well presented to the customer?), sales material (flyers, product boxes, stands) and constant maintenance of the sales relationships. Often sales contests and bonuses can be included to teach the distributor to sell the service or to increase its market success.

To make sure that the service is sold properly, the distributor has to profit from turnover more than just provision. As soon as the distributor's interests and the service provider's interests meet, the relationship is prosperous. This might be in simple areas, such as providing compelling sales argumentation, or more complex activities such as sales campaigns to boost demand. The ideal sales representative is one who remembers to offer actively a service and recognizes the right target segment. The sales representative cannot be expected to discover the best mix of arguments to segments. The service owner(s) need clearly to define the target segments and offer sales argumentation in the training for selling the service.

It is a hard game to differentiate your 3G service from those of other 3G service providers in the distributor's heart and mind. Not only the 'desire'-marketing of the service is important, but so is the closeness of the provider and the distributor. Even though little promotional gifts are nice to have,

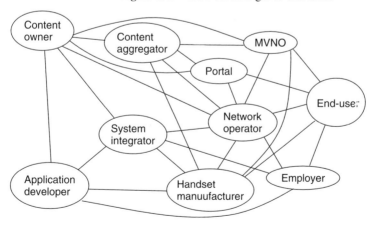

Figure 13.1 How services get to end-users

the flow of information and extent of support play a bigger role in success in the long run. Operators will have to be even more imaginative and innovative in the choice of their distributors. Totally new possibilities can come up depending on the service in question. Positioning services and wilderness trecking equipment stores fit together and 3G services for hiking can be offered there. Similarly 3G adult services and sex shops fit together. As the 3G service portfolio expands, soon customers will be buying 3G services offered at the hairdresser's or at the supermarket. Operators can tie sought-after distributors by exclusive deals and higher revenues. Two strong brands on the operator and distributor side might profit from the synergy and brand enforcing effects.

13.3 Selling to consumers

Selling to consumers has historically been via mass distribution by mass communication methods. Even with 3G the majority of the sales effort will be done through mass-marketing methods of advertising, etc. Still, sales to consumers will become an increasing opportunity in 3G as services are created much more for special segments. Therefore selling 3G products to consumers is very much tied to segmenting the sales channels too.

Selling 3G to consumers is roughly a two-step process. First, the consumer has to become a 3G client, i.e. he has to acquire a 3G terminal and enter a subscription relationship with a provider. Here we face very much conventional selling via specialized 3G dealers such as the network operator's stores, independent mobile phone stores and electronics outlets, etc. The issues involved were discussed in the distribution chapter.

The second phase, however, is selling different add-on services to the customer who already has a subscription or is in the process of acquiring a subscription. The phone dealer plays a crucial role, again, but service sellers could be any kind of store from the news-stand or kiosk selling postcards (and 3G picture messages) to the locksmith selling locks (and 3G security services). It is likely that 3G itself, i.e. mobile portals and the direct contact with the subscriber, will be among the most effective sales channels, because of the billing possibilities of 3G.

13.3.1 Event related sales

Sales opportunities can have an event dimension with, for example, some youth-oriented services sold at a skateboarding or snowboarding event, where an operator could set up a booth to sell and provision youth-oriented services such as messaging, chatting and dating services. Naturally the staff at the booth needs to be appealing to the target customer — youthful as well in this example — and very importantly the services sold must be able to be activated immediately.

13.3.2 Bundling services with the subscription

Initially, many of the early 3G services can be sold as bundles with the subscription. These could include selling services such as daily news, video clips, horoscopes, etc. The initial subscription service could include a basic data-services bundle with for example an information package, a selection of games, a set amount of free SMS and MMS messages, etc. Additional use of further services would then be billable according to set rates.

13.3.3 Billing inserts

One of the powerful tools the network operator has to sell its new services is the monthly bill sent to the home of the residential customer. This gives an opportunity to mention services in the bill itself, and to include various advertisements and brochures as inserts in the billing envelope. As many 3G services can be of a seasonal or fad-like nature, the monthly bill can act as a powerful means of activating users. For example, the upcoming major sports event such as a Formula One race or football championship, etc., can be used around launching a related service. Another example is annual calendar-related events such as Mother's Day, Christmas, etc. These too can be used to introduce new services allowing subscribers to share the celebration with each other.

13.3.4 Portal placement

A new opportunity for the mobile operator and the mobile portal owner is direct placement on the portal. This may be in the form of a welcoming message at the opening screen of the portal, or be linked to any given page — such as seeing a new service for sports fans advertised only on the related sports pages. What makes the portal a potentially superior sales channel, is that it can be mass-tailored to suit the individual's needs and tastes, and the portal allows for immediate digital and 3G network-based service provision. The service desired can be immediately activated from a 'sign-me-up' button or 'click-to-buy' link. For portal placement to work as an effective sales channel, however, the operator has to get its users to use the portal frequently. The early feelings relating to WAP and GPRS portals in Europe provide a poor starting point from which 3G operators train their users to love the 3G portal and to use it often. The information from Korea and Japan suggests that this goal is quite attainable, however.

13.3.5 Selling to businesses

As with residential customers, so too with business customers there are two phases, selling the subscription and selling further services. Business-to-business (B2B) selling is a more complex process. The linchpin of successful B2B sales is the account manager. While selling to consumers involves at least some economies of scale, and sales efforts can be enhanced with mass marketing efforts, selling to businesses is almost always a personal and a highly committed activity. Even though sales turnover figures are much higher in B2B, profitability is diminished by higher sales efforts, production costs (especially of tailor-made services) and discounts granted. The business customer segment is a heavily contested market, especially with large corporations. This means that even after winning a contract, the competition has driven the profit margins to minimum acceptable levels. In fact, often the largest corporations can extract price levels that can be unprofitable in the short run.

A vital personal attribute of the sales representative when serving business customers, is the ability to stay calm. With businesses adopting new services whether from you or from other suppliers, inevitably there will be problems. Sometimes they may be the fault of your company, sometimes that of the customer, and quite often the result of some third party, such as an outsourced IT support company, etc. More often than not, your company will get the initial blame. If the telecoms system is partly involved, it will soon be blamed. Inevitably the sales rep will get the call and the initial blame.

13.3.6 Corporate customers

To understand how business customers differ, we can start by examining them by size. Remembering what we said about segmentation, and that grouping by size is no more efficient from a marketing segmentation basis, we will still simplify matters in this chapter by dividing business customers into the traditional three sizes. They are large corporate customers, SME (small and medium enterprise) customers and SOHO (small office–home office) customers. Even thought the biggest and best known large corporate customers are very attractive as reference customers and serve the PR interests of the 3G service provider, the large corporate customers seldom surrender particularly profitable deals from their suppliers. 3G-services providers may be so keen to get the 'big 50' onto their customer reference lists that big businesses can usually play them off against each other. In addition, large corporate customers require much attention from a large sales staff and customizing work, including integration with numerous IT systems. Many of the larger government units may be similar to the large corporate customers in how they approach telecoms services.

13.3.7 Large corporate customers

Depending on the country there might be anywhere from 20 to 200 large corporate customers. The big and even many mid-sized corporate customers will tend to make their telecoms and IT project decisions on two main concerns. They will expect an 'ROI' or return on investment calculation and justification for the project. The ROI will involve the costs of the needed technology and its various integration, implementation, training, etc., costs, which are compared against the expected savings or additional revenues that the project would produce. An ROI analysis will usually also include an estimate on the payback period of the project, measured typically in months. These customers often have dedicated telecoms specialists who, surprisingly, are often not closely involved with the IT department and may be in strong internal conflict with the often more powerful IT department. Also, large corporate customers will not want to disrupt their IT support functions. Large corporate customers tend to be conservative in their technology adoption and will not want to be the testing grounds for new ideas.

13.3.8 SME or medium sized companies

SME customers are, naturally, smaller than large corporate customers but there are many more of them. Typically, SME customers number in the thousands. There is a very wide variety of behaviour in SME customers as we discussed in Chapter 3, and this amount of business customers contains numerous attractive segments. SMEs often are in single facilities or offices, or can be local in reach such as a chain of restaurants. SMEs tend to have

Figure 13.2 Complexity in service creation (theory by Merja Vane-Tempest)

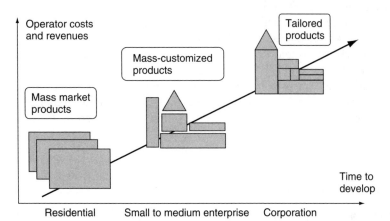

specialists who take care of IT matters but rarely have dedicated telecoms staff. Often it is necessary for the SME salesforce to be 'fluent' in the vocabulary of IT, and knowledgable in the translation of telecoms terminology to their IT equivalents. SMEs have a wide variety of corporate cultures, and in any given industry there may be some that are ultraconservative and others who are extremely innovative. National new technology innovators with business solutions tend to be found in SME size companies.

13.3.9 SOHO or small businesses

The SOHO customers, or the many small-sized companies and one-man trade businesses and sole proprietorships, are in many ways similar to individual consumers. With the smallest customers the owner tends to be the decision maker and has to handle all decisions including those relating to telecoms matters. The owner is very unlikely to want to spend much time with understanding the finer points of given 3G services. The lack of time is a big hurdle for often technophobic small businesses. Major sales arguments for small enterprises are therefore ease of use, cost efficiency, and process improvement.

We need to point out that the above categorization suggests differences in selling to business customers by size. This is not a meaningful segmentation for developing services, understanding customers or for developing sales argumentation. Such activities will need to be well developed and could not possibly be given fair treatment in half a dozen pages in a chapter on sales. Please see Chapter 3 (Segmentation) for more information.

13.4 Selling to partners

Selling to partners is a new phenomenon for mobile operators. Up to the late 1990s, mobile operators only sold in the direction of the end-user, whether directly or through channels. With new mobile services, especially with 3G, the opportunity arises to sell 3G services to partners including content providers, application developers, IT integrators, the advertising and m-Commerce players, etc.

Partners are approached by an own-channel sales crew and need special care just as the distributors do. With most 3G network operators (or MVNOs) and/or mobile portals, a partnership manager will be set up for the most significant partners. These will include, for example, the banking partners, advertising agencies, news services, serial content providers such as cartoons and horoscopes, etc. The partnership manager will be a liaison between the 3G service provider — network operator, MVNO or mobile portal — and designated partners. The partnership manager will probably not be able to discover independently all possible services that his/her employer can provide to the partner. There will need to be service managers to build and manage services tailored to the partners, and in many cases some kind of sales representative will be needed to turn a service opportunity into a contract with a partner. Selling to partners may be seen as a VIP service.

Services provided to the partners include those relating to the natural strengths of the network operator, and where those are provided to the MVNO or mobile portal. These include location information, the status and/or presence information of the 3G terminal or its user, usage information by customers and by services; as well as charging and billing services; provisioning; branding and co-branding; and placement on the portal. Depending on the strengths of the partner involved, the services sold can include many others including IT integration work, design, IP services, credit risk, etc.

The actual services provided, their tariffing, etc., will depend on the skills and interests of the provider and the competitive environment. For example, one network provider may want to provide a broad range of credit, money transfer and fee collection services in addition to charging and billing. This type of network operator will probably need to get a banking licence and will be seen as a new-entrant banking player by other partners in

Figure 13.3 How location-based services relate to human needs

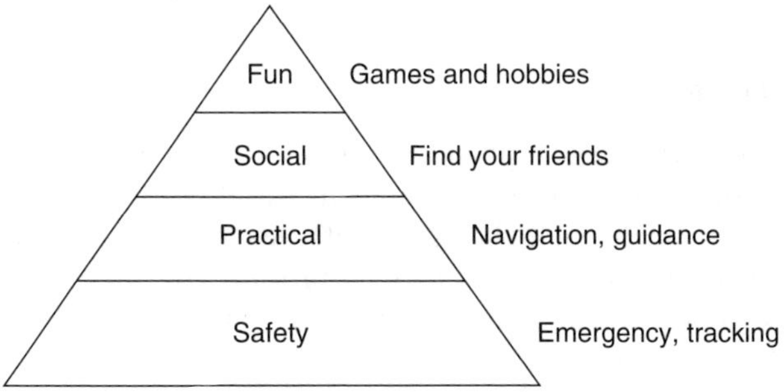

```
            /\
           /Fun\      Games and hobbies
          /------\
         / Social \   Find your friends
        /----------\
       / Practical  \  Navigation, guidance
      /--------------\
     /     Safety     \ Emergency, tracking
    /------------------\
```

Source: TEKES National Technology Agency of Finland, 2003

m-Commerce. Another network operator may want to take a minimal role in only charging and billing, and get a banking partner or partners to handle the more advanced banking services. We discussed partnerships in greater detail in Chapter 5.

13.5 Motivating the sales representative

As we discussed briefly in Chapter 7, a significant key to how mobile phones, accessories, subscriptions and services sell is the actual sales representative. Even though the sales reps are often burdened with inconsistent or even conflicting goals and objectives, they do want to serve their customers. The sales representative will need essentially four vital things. First, the sales rep needs good information on what is to be sold. The better the sales rep knows the service or product intended to be sold, the better the job can be done. Second is the motivation, in that the sales rep gets sales commissions, sales contest awards, and bonus targets. Third, the sales rep needs access to the customer, which in consumer situations means creating pull by advertising and promotion to bring the sales prospect to the store or into the calling centre. The last and often neglected part is that the sales rep is also a passionate user. All the services that a sales rep is intended to sell should be made available for free to the in-store sales staff. They should be encouraged to use the 3G services to the fullest, to become evangelists. And of course there needs to be good sales support materials.

If the sales reps are not supported in these ways they will be demoralized. Typically, it can take very little time to sell a service to a new customer but much more time to teach them how the service is used. The sales reps need to be motivated to spend the extra time, otherwise the customer might never be comfortable using the service. Or worse yet, the frustrated user will call up the help desk and then training the user can take much more time than if the sales rep had done it when being with the customer.

Of course before any sales rep can start to train *others* to use a given service, they must be taught themselves. As 3G will be evolving fast, the sales reps will need a lot of training in advance of various service launches. Both in-store sales staff and any sales reps that visit customers should be given self-study courses, incentives and time to complete those courses and develop their know-how.

Sales reps are very predictable and any sales plans and targets should be run by a small trusted sample of current senior sales reps who know the storefront situation. These sales reps should not be punished for giving their own opinions. This test group should not be burdened with the simultaneous task of trying to re-educate the other sales reps, as that will never happen. A sales rep is driven by his greed and the perceived value of the various benefits that are available. The reason for the small test sample of senior sales reps is only to ensure that the average sales rep will react to the intended campaign as management wishes. The opinion of these experts should be taken very seriously and they should not be criticized for voicing any objections. Whoever designed the sales campaign should be advised that at this stage the campaign *will be revised*, based on the feedback from the test group. That ensures that the design of the campaign is not made so rigid as to prevent changes. A quick sanity check with the trusted senior sales reps will save a lot of headaches and help to ensure that any new promotions, sales plans and targets can reasonably be met.

One cannot overstate the importance of getting the sales reps to use the services they sell personally. It is a very different occasion for the prospective customer when another user shows a new service, rather than only a disinterested sales rep. As every sales rep cannot be expected to be made an expert on every one of thousands of new 3G services, the sales managers should monitor the interests and try to find suitable groupings so that in any given situation if a customer is interested in something specific, a specialist using that service will be available. The sales reps can do such things as set up hyperlinks to relevant pages and help with the early settings on the customer's actual 3G phone when doing so, again ensuring that the customer will rapidly be up and using 3G services.

13.6 Handset subsidies

Handset subsidies are a peculiar aspect of the cellular telecoms industry. They exist in most countries and can be quite substantial. Where subsidies may have had strong business cases supporting their introduction 10 years or so ago, the subsidy issue has changed considerably and we feel compelled to discuss the subsidies at this stage.

Handset subsidies are common in most markets and many analysts have attributed the rapid adoption of mobile phones in part to subsidies. It may surprise readers to find out that Finland and Italy — which have consistently been among the hightest mobile phone penetration countries and by early 2003 had exceeded 90% — have no subsidies as they are illegal. The cases of Finland and Italy prove that world-leading mobile phone penetrations rates are quite feasible even when the subscriber has to pay the full value of the handsets up front upon signing for the service. We must conclude that handset subsidies are not necessary for the industry.

However, are subsidies beneficial? As we said, in most countries subsidies are part of mobile telecoms. For example, in Germany 2G subscriptions are tied to handset sales. Operators subsidize mobile phones by tying customers to contracts of a certain length of time, usually from 1–2 years. This practice has produced several unwanted and unanticipated effects.

At first glance, one might make the hasty generalization that low terminal costs reduce the barrier for customers to enter the 3G market. As we saw from Finland and Italy, it is not necessary for that, but handset subsidies are a very short-sighted tool for the network operator, since for the network operator it is not the terminal costs that are crucial but rather the use of 3G services. Subsidies can damage the business opportunities for 3G services. Since handset subisidies usually go hand-in-hand with higher usage fees — to recoup the subsidy — the ratio is out of balance. The network operator should sell its own services at market rates, not above them. It is folly to subsidize the handset, where the profit goes mostly to the handset manufacturer, at the cost of fewer services consumed because of artificially high service prices.

As customers have grown familiar with the idea of subsidies in most markets, they have also learned to extract the best benefits from this system, effectively abusing the system. It is common that whenever a subscriber wants a newer phone, they will go to a competing network operator and sign up on a new subscription to get the phone. Users have 'lost' phones by throwing them into the river, etc., to get new ones. The further the handset

Figure 13.4 Worldwide handset market shares 2003 (IDC November 2003)

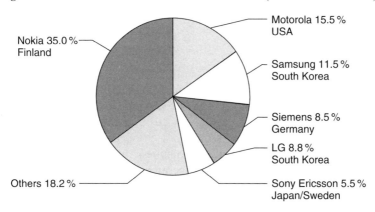

price is from the real market price, the more the user is enticed to try this type of trick and gimmick. What subsidies have done is destroyed rational consumer behaviour and in its place created a mindset of artificially high churn.

There are other side effects of subsidies. Handset manufacturers are deprived of an important part of brand creation: price. Customers do not like to be tied to an operator with minimum contract periods. Customers should have the freedom to change the operator easily if they like.

Of course, we have to admit that handset subsidies might shorten the terminal product life cycle, enable rapid initial adoption, and give opportunities to distribute new terminals with new features in an even faster time frame, but again, it is not the terminal prices that matter in 3G as much as the service prices.

Because of their considerable harmful effects to the market, handset subsidies have generally been found to be undesirable by most experts commenting on mobile telecommunications around the world. The network operators are trying to find ways to diminish and eventually extinguish subsidies. In Korea, the operators went so far as to lobby the regulator to make subsidies illegal. This was done and subsidies were removed from the Korean market. In most markets subsidies present a traditional prisoner's dilemma, i.e. where behaving alone for individual gain goes against the greater gain for all of behaving as all others. The problem lies partly in the long-standing tradition that has been established. In most markets, customers have been trained to expect subsidies and may react negatively if subsidies are suddenly removed. The current thinking, therefore, is to try to diminish the sizes of the subsidies gradually.

13.7 Non-traditional sales

We predict that sales will become a more significant factor with 3G than it was with 2G. This is because services are becoming increasingly interchangeable and mainstream mass-market services ever more homogeneous due to the greater degree of standardization, co-operation and partnerships. Consumers will show less operator stickiness and be more open to buy services from providers other than the operator whose subscription they have.

13.7.1 Cross-selling

One of the very important methods is cross-selling. One of the natural opportunities for cross-selling is the operator portal. Already today on the fixed Internet, portals are the most important cross-selling channels for many Internet service providers. This will be even more the case for mobile portals. On the other hand, mobile portals do not offer similar advertising possibilities to current advertising via desktop computers, since ads on the mobile portal are perceived as annoying taking up prohibitively large proportions of the small terminal displays. However, sponsoring a part of some services gives opportunities to promote others. Strong brands along with excellent, attractive services play an equally big role in cross-selling.

Another opportunity for cross-selling is whenever the customer (subscriber) contacts customer service, for example regarding billing issues. The customer service representatives will need to become sales representatives at the same time. They need immediate real time access to the customer's usage data and intelligent sales tools to assist in rapidly identifying their segment, and likely interests and needs. For example, if a mother calls to enquire about putting limiting controls on the child's mobile phone, the sales representative can offer to sell her a family calling circle bundle for all mobile phone users, perhaps even converting another child's subscription on a different network, to the one with the family plan. Decisions on opportunities have to be made fast, but every customer contact with the operator is also a chance for cross-selling new services. With the rate of new service launches in 3G, even if the customer calls every month, there always can be new services to sell.

Figure 13.5 Global mobile gaming revenues: $1.5 billion in 2003. Mobile gaming tripled in size from 2002 and generated 4.4% of the worldwide gaming industry revenues (The Research Room July 2003)

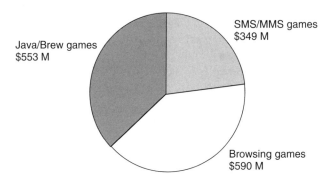

13.7.2 Bonus point programmes

Cross-selling can also be achieved by bonus point programs. Many operators have introduced bonus point schemes similar to those by the airlines awarding gifts and rewards from a variety goods and services provided by partners. The benefit of rewarding customers with bonus point programmes is that the operator's own services can be awarded at the face value while the actual cost to the operator is naturally the wholesale cost of the service. Customers will feel as if they are receiving greater value than what is the strict monetary value of the service. Bonus programmes can be used to allow customers to earn handset upgrades and replacements.

Some may feel that bonus programmes present a strong threat for the customers to leave once they have used up the current collected set of bonus points for some award. Recent experience has, however, shown that customers tend to feel very loyal around the time of redeeming rewards and there is no reason to expect that the customers would churn soon after that time.

13.7.3 Network effect/viral selling

The best promotion for products is always your own confident customers. Many of the most successful mobile services, like ringing tones and logos, are sold through word of mouth in networks of people. Obviously, the one who controls clans and communities via services has a perfect sales channel working autonomously and often even more passionately than the best of sales forces. 2G services offer, for example, special phone or SMS fees for registered groups of people like friends or families. This creates peer pressure to get the same subscription, 3G terminal and services as most of your friends. Connecting friends and creating clans is a big business itself and will create powerful upcoming sales opportunities. The first players on the market will therefore have a certain advantage in forming communities.

Network selling is achieved by concrete 'member get member' campaigns. The newer a technology is, the more the customers have to be educated and taught to use a new service and this can be done best and most economically by having customers tell their friends how to do it. The targets of initial sales efforts should be the early adopters in these groups. More about the way early adopters can influence the success of any services in general was in Chapter 11.

13.8 Sales out

This chapter discussed sales of abstract services and the soul of the sales representative. Note that many of the issues are closely linked to those involved in the distribution channel (Chapter 7). No matter how wonderful the new 3G services are, if they are not sold, the services will not have a chance to deliver satisfaction to their intended users. The sales people and processes are vital to 3G success. Increasingly the line becomes blurred between what is sales and what is anything else within a business entity. As Jeff Woodruff said: 'Everybody works for the sales department'.

Look at me: I worked my way from nothing to extreme poverty.
— Groucho Marx

14

Tariffing
Pricing for Profit

This chapter looks at tariffing for mobile services. Tariffing determines the profitability of the whole business, and has a bigger impact on it than do network dimensioning, traffic optimization, sales provisions, handset subsidies, or any other decisions made relating to mobile services. The basic principle relating to tariffing is that not all customers are willing to pay the same price. A major determinant of how much price matters is who pays for the price of the call or the mobile service. Typically, when services are paid for by an employer users are much less sensitive to the price than if they have to pay for the service themselves.

3G Marketing: Communities and Strategic Partnerships Tomi T. Ahonen, Timo Kasper and Sara Melkko
© 2004 John Wiley & Sons, Ltd ISBN: 0-470-85100-7

14.1 But isn't tariffing simple?

Tariffing is an area of mobile services marketing that mobile operators *think* they know well. What is there to it? Run a survey of what competitors offer or a survey of what customers are willing to pay, and set the price. Simple? This kind of thinking abandons dramatic amounts of profits and alienates countless prospective customers. Modern scientific tariffing is an intergral part of the marketing mix and is founded on the principle that some customers are willing to pay more for the same service. If the basic service is already priced to cover costs, then that increased pricing opportunity is pure profit.

Still, today, with most telecoms operators, tariffing is often an afterthought, or left to a novice product manager fresh out of university. Worse, pricing may be a relic of history resting on concepts of some service idea from long ago, possibly tracing its history to monopolistic times. Tariffing directly affects the profitability of the company. That is why this chapter is needed. The reader will need to understand such issues as price elasticities, opportunity cost and pain threshold. Before we get into those, let us start with basic fixed goods tariffing.

14.1.1 Cost-plus tariffing

Normally, when it comes to traditional, tangible products, the price is calculated by summing up production costs (raw material, personnel), a certain share of machine investments and other costs, and then adding a certain value-added profit margin on top of it. Of course, the price cannot be higher than the customer is willing to pay, but the value could be at least partially calculated and measured by estimating costs of the materials that were used during the production process.

With abstract services, the calculation gets much more complicated and means that working hours of the employees have to be measured and given a certain value. With digital services there are often large development costs, but almost no incremental costs of selling multiple copies. For example, games are likely to experience this phenomenon too, where the gaming logic has a given development cost and of course marketing expenses, but once developed it does not cost much more whether the game is sold twice or sold 5 million times. How do we account for such costs in pricing? Cost-plus tariffing can be used in manufactured goods and even in many service industries, but with telecoms and advanced wireless services it is a complicated and imprecise mechanism. What is more important is that tariffing should be based on what customers are willing to pay and what market pressures allow. The only other consideration is, of course, that unless expressly so desired for some marketing reason, services generally should not be offered at a loss.

14.2 Some customers are willing to spend more

The underlying premise for tariffing and segmentation is that some customers are willing to spend more money for a similar, sometimes identical, service than others. For a service provider in any industry the ability to identify which customers are willing to pay more is a key to profitability. The company will need to develop services, service categories, service bundles, targeted tariffs, targeted marketing campaigns, etc., to be able to capitalize on the differences in willingness to pay.

The need to fill capacity is even more important in a capacity-driven industry such as telecommunications. The network is built and has a certain capacity, whether nobody places any calls, or all capacity is used. There are very few chances of variance in costs in operating the network, so maximizing usage of the current capacity with the most profitable services and extracting maximum price for those services is the intention.

14.2.1 Airline analogy

There are a lot of similaritites between the airline industry and telecommunications. Both involve global transport: the airlines mostly of people and parcels; telecoms mostly voice and data. Both build networks with major routes and hubs. At nearly 100 years of age, air travel is the younger of the two industries. Both industries initially had national monopolies. Both are seeing increased competition and a commoditization of their main products. The airlines have had experience with competitive pressures for much longer than most telecoms operators, and so many lessons can be learned from airlines about product development, route optimization, customer segmentation and loyalty programmes and then tariffing strategies.

Lets examine the key pricing lesson through an example. If Lufthansa has scheduled a flight from Frankfurt to Chicago, it will most often have to fly the plane whether that particular flight is sold to 95% capacity or to 15% capacity. The odds are that there is also a return flight for that very aircraft with seats sold from Chicago back to Frankfurt, and so even if one flight

Figure 14.1 Typical airline cabin classification

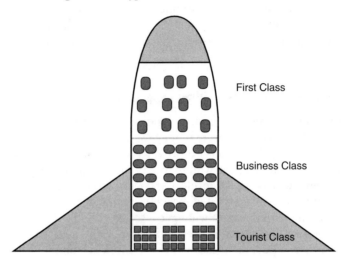

happens to be near empty, the chances are that its return flight is more than half full. Thus planes can fly individual legs of flights with relatively few passengers on board simply because the route requires it and the complete round trip journey has enough passengers to justify the trip. Naturally airlines are also not eager to cancel flights that cause inconvenience to their passengers.

Lufthansa will need to fly the route and the need therefore, is to fill the planes, and to try to get as high a price per passenger as possible. This is why there are several flying classes, such as first class, business class and tourist class. To the casual viewer, and thus most passengers, this seems like a very simple pricing plan, with three clearly defined pricing classes and a clear offering of what you get for your money. Tourist class gets tight seating and limited food offerings. Business class gets better seats and food, and also gets to board and disembark the plane before tourists. First class gets very good service, full-bed reclining seats, etc. It all seems simple and reasonably 'fair'.

Actually, if you examine the prices paid per seat in any of the classes, there is huge variation where one passenger might very well pay over double for an identical 'class' seat than another. In fact, probably on most flights, the cheapest fare paid by someone in business class often is less than the highest fare paid by someone in tourist class! How is that possible?

The price of airline tickets will vary not only with class of seat (class of service), but also with several other elements that tend to be hidden from the purchaser. First of all there is time — usually if you buy very much before the flight, such as 3 weeks before, you get cheaper prices than if bought on the week of the flight. And if you buy on the day of the flight you will pay even more. Some prices are promotions that have limited seat quotas.

The price varies by location of purchase, typically the same Frankfurt – Chicago – Frankfurt round-trip will have a different price if bought in Germany (departure point) or in the USA (the destination of the trip). In fact, the same round trip will have a third price if that same fare is bought from elsewhere such as the Lufthansa office in Paris. There is reason for some differences in prices due to currency variations, but airlines are very clever to use this as the convenient explanation, while the actual differences are far greater than what currency exchange-rate fluctuations would allow. The airlines have learned to offer competitive rates on competitive routes that are only dependent on the competition of the purchasing location.

The price varies by bundle — if you buy a Lufthansa ticket from Munich via Frankfurt to Chicago, the total cost will typically be less than if you bought one Lufthansa ticket from Munich to Frankfurt, and another Lufthansa ticket from Frankfurt to Chicago.

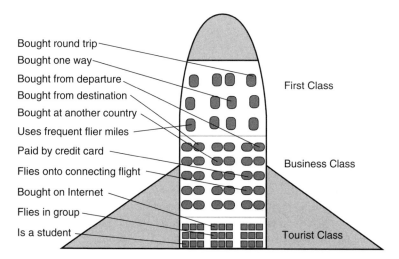

Figure 14.2 Actual airline seat-price variations

The ticket price varies further for reasons such as distribution channel (different travel agencies may charge different prices) and purchase method (Internet purchases are often at a discount), sometimes by payment method (a credit card company may have a pricing special), or by customer affiliation (membership for example in a tourist association or citizens group may entitle a discount). Seats are sold also on bulk discounts (group fares, tourist agencies use this method to fly groups of travellers to popular destinations). Then there are the wholesale clearance resellers, who fill seats. These include all kinds of innovations such as bids and auctions. And then there are prices customized by customer in other words some large corporate customers get special discounts. The airlines offer special prices and deals to their frequent fliers, offer occasional free upgrades (hence *defacto* discounts), and free trips after the passenger has flown a specified number of miles, etc.

The number of different tariffs for one route may seem mind-boggling. When the full analysis is done, a typical major airline has thousands of tariffs for every single route. They have not emerged by accident but rather by a systematic segmentation process of fine-tuning every separate ticket sales opportunity with the exact conditions that apply in order to extract the best price possible. Thus behind every pricing decision is a lot of customer understanding and the aim is constantly to maximize the revenue per seat, and hence profit, on any route. Incidentally, low-cost airlines that promote themselves with simple and cheap prices, actually have very similar systems. The seat prices for any route vary very greatly according to similar rules. The only major difference is that low cost airlines do not offer business and first class travel. This helps create the illusion that their pricing is simpler and somehow 'more fair' than those of the full-cost airlines.

The consumer insight behind such pricing plans is considerable and has evolved over time. If one person needs to fly to a given destination, and knows it 2 weeks before, and another needs to fly to the same destination, but only finds out 2 days before, it does make sense to 'punish' the late person by forcing that person to pay more. The airline is in the business of filling seats. The chances are that most alternative and possibly cheaper seats are already sold. To protect the tariffs, the airlines have incorporated numerous rules into their special prices, such as the Saturday overnight stay at the destination. The difference between a regular tourist ticket (which few passengers actually pay) and a discount tourist ticket (which most tourist class tickets tend to be) is the rule necessitating a Saturday stay. This balances loads away from the most frequent days of travel.

14.2.2 Applying the example to telecoms

Can this principle be used in telecoms? Of course it can, and in small ways it is already being done today. Let us imagine that you are travelling to Singapore and need to contact a friend visiting Hong Kong. First let us assume you are a German businessman and your employer pays for all of your mobile phone calling costs, including personal calls. You would have no problem calling your friend in Hong Kong. If your employer does not allow you to place personal international calls, then you would consider the cost of calling from the hotel room — and perhaps consult the hotel calling guide to see how expensive it is. Depending on your propensity to spend and the need of the contact you might call, or you might for example send an e-mail message to your friend rather than calling. You probably would not bother to search for the lowest cost calling cards, i.e. you would essentially consider the relatively expensive hotel room call 'worth the contact'. You probably would not consider renting a mobile phone to be able to place the call. You might consider going to the payphone and perhaps using your credit card or coins to pay for the call. The rates for coin calls from a phone booth would often be less than the price from the hotel, while the call might be handled by the same telecoms operator. So already here we have the same voice minutes from Singapore to Hong Kong charged to the same person at different rates.

In the same market there are numerous operators who offer calling card services with which one can place much cheaper calls. If you were living in Singapore, or visited it often, you probably would get such a calling card. Now we have several call prices for the same route. Actually there is a wide variety of other pricing elements, such as congestion — calls during office hours tend to be more expensive than during the evening — and of discounts for contract customers, etc. In the above example, we assumed that the calling person and his condition is the main element for allowing change of tariff. Actually, the reason for the call may be a more significant element. If we change the picture just a little bit, in that the person in Hong Kong is your boss, he is about to make a decision on your yearly bonus, and has asked you to call on this day, then the price of the call will no longer be an issue. You would be willing to even to rent a mobile phone just to be able to call your boss, rather than get less of a bonus than you feel you deserved.

The above example illustrates both that telecoms operators already use variety in pricing a similar or even same service, and that users will have different willingness to pay, and that willingness may even change per person depending on conditions.

To understand who the customers are, what their needs are, how your services can address those needs, and what those customers are willing to pay for your services, is vital for ensuring that you maximize your profit. Some customers in some situations are willing to pay a lot, and you should make sure that those who are willing to pay the most will always have easy access to your services. Other customers are not willing to pay as much, and you should have various flexible means to maximize your traffic and utilization, and also to make sure that those paying less will not crowd out those willing to pay more.

14.3 Profit and pricing

When simplified, profit equals sales price minus costs. It may seem very simple to say that you can increase your profit by raising your price, or by lowering your costs. The thought of raising prices easily raises great concern among telecoms marketing and pricing managers and product managers. It should not. Price fluctuation is a natural phenomenon in economics, and telecommunications has been quite exceptional in that most prices have been remarkably stable over decades.

As mobile telecommunications starts to follow common business laws more closely, so too will issues like price fluctuations become ordinary everyday occurrences. With any given product or service there is likely to be a zone of inertia, where a small change one way or another is not likely to cause measurable immediate change in consumer behaviour. If the difference is too small for users to care, it will not matter. With any given price change, a very significant factor is the perceived price differential to perceived competitor prices.

It is important to note both instances of *perceptions*. The absolute *truth* is irrelevant. If the customer perceives that one competitor's prices are lower than yours, that is the only relevant truth in your market, until you or some other factor in the marketplace change that perception. Equally, the real price differential may be meaningless or tremendous in size, that too is irrelevant and only the customer-perceived price differential will matter. However, that matter, together with the price elasticity, will determine how many, if any, customers will change behaviour because of a change in pricing.

Figure 14.3 UK price fluctuations during one year (OFTEL August 2003)

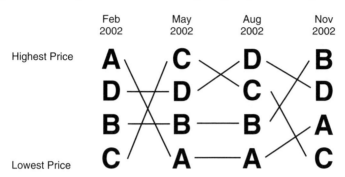

UK regulator OFTEL ranked post-paid mobile prices in UK including cost of handset, without naming names reported the rankings.

Note that behaviour change can be at several levels. Let us assume that you raise your price of calls for a given segment by 5%. That could result in your customers in that segment reducing calling minutes by 5% (more or less). It could result in your customers shifting calling to another time or using another service, such as sending more SMS text messages. It could result in in some of your customers ceasing to use of the mobile phone as a type of protest for the price increase, or it could result in some of your customers churning to your competitor partially or wholly. There may also be word-of-mouth effects, where a happy customer may stop promoting your service, or an unsatisfied customer starts to tell all friends what all is wrong with your service.

All of those effects are only in the relationship between your company and your customer. The customer, however, does not view your price change in isolation, but rather in the context of your competitors and other services. Any changes to your prices will be viewed in the context of what happens in the market. The worst time for you to raise a price is when the perceived competitors have reduced theirs. The best time to raise prices is when other related prices are also increasing. The follower in raising prices usually benefits the most, the first to raise prices suffers the worst market reaction. Conversely the first to reduce prices usually gains market acceptance, while others, even if they reduce prices more, tend to be considered as mere followers, responding to the lead of the price leader.

14.3.1 Prices and usage

What is also comforting is to see that the laws of price and demand seem to apply in a common-sense way, that roughly in proportion to prices falling, mobile telecoms usage has been growing. This was documented for example by Deutsche Bank in its *Wireless Internet* report when examining the UK mobile telecoms market and the advent of prepaid customers. With the overall drop in prices, usage kept going up, keeping total revenues in proportion about the same.

While voice tariffs can be expected to follow existing prices and adjust gradually, new services have no such past history and pricing has to be set 'in the dark'. The problems start with getting the initial pricing wrong, pricing too low, and then, if forced to raise prices, customers will be driven to your competition. The risk of pricing too high is that nobody will use the service, and repeated lowering prices may be seen as an admission that the service is no good anyway. With new mobile services the issues get even more complicated as we are not looking at only one service. Mobile operators and their service providers have to consider each service separately for tariff and profit.

Figure 14.4 Minutes of Use (MOU) overall have remained relatively stable, with prepaid use and postpaid use both growing to compensate for the increasing proportion of prepaid accounts

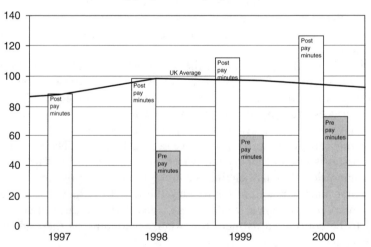

Source: Deutsche Bank Wireless Internet Report May 2001

Each service needs to be considered separately by looking at its competition, and examining the price elasticity for that service. With pricing there are two costs: if the price is set too high, there is the cost of losing customers; if the price is set too low, there is the probable cost of lost profits. We have already seen from the airline pricing example that different consumers are willing to pay different amounts, and even the same consumers may be willing to spend different amounts depending on the conditions. This concept needs to be applied across all services with tariff-optimization held in an ever-changing competitive landscape.

We believe that most readers of this book will feel very uncomfortable with the idea of offering different prices to their customers. Keep in mind that we are not talking about promoting the fact that there are differences in how customers are treated. Of course customers should have a feeling that they are special, and also they should never get a strong feeling that some customers are 'more special' and receiving a better price. Recognize that already today mobile operators are doing this. A prepaid residential customer will be paying a different price per minute and for SMS than a large corporate customer who has 5000 subscribers and has negotiated a corporate discount on its total spend. We already are in the world of differences in prices, it is now only a matter of doing it more systematically for profit.

14.3.2 The variety in acceptable price

The prices of products or services which have a long market history tend to achieve a zone of acceptable levels for the price. In other words, prices harmonize. For example, the price expected for a carton of milk or a haircut is pretty similar in any given market. One of the reasons for this is the efficiency of markets in creating sustainable price levels, where customers will compare prices and those who price too high will lose customers, while those who price too low go out of business. In telecoms, one could argue that the calls of traditional fixed telecommunications tend to fit this pattern. In most open markets the price variety between major players, say, of prices to the most common international call destinations, tend to be similar. Over time, tariffs for homogeneous products tend to harmonize, as has been shown by the UK regulator OFTEL in its study of the average price of mobile call minutes in the UK market.

With new services no such history exists. The first to the market has considerable freedom in setting prices and if the service proposition is compelling, there will be customers who will pay almost any price. The problem with mobile services is that their profitability mostly depends on mass adoption. If the price is set too high, small revenue streams will trickle in, but the large revenues and profit from mass market economics will fail to materialize. The initial data from usage suggests a high profit per customer using the service, which may perpetuate the problem. The service never takes off in the mass market.

The temptation with new services is to price at the higher end of early market research. The conventional wisdom of much management is that it is much easier to reduce a price than it is to increase it. The temptation to price at the high end should be resisted at all costs. Remember that marketing is hard work and we are not in this business just to do things the easiest possible way. There are numerous ways to make customers accept increases in prices, such as initially offering clearly defined 'trial pricing', and also building bundles and service improvements. Any of these can allow for satisfied customers to accept an increase in the price of a service.

The problem with pricing at the high end of a new service is that it almost certainly fails. It fails to build a mass market for the operator who introduced the price, but it opens the door for a competitor to capture the existing customers *and* the mass market. If one player plays the high price game, educating the market but not capturing it, another player can build the mass-market version and take the whole market, including most who initially used the expensive and exclusive service. As service creation will be in the digital IP environment, no service can technically remain exclusive for long. Competitors will come, and if the first player does not capture the mass — or at least its fair share of the mass market — it may never do so.

14.3.3 Prices for service introduction

What level to set for launch? Usually some kind of market research is used to determine what customers might be willing to pay. That usually results in a range of prices and the common way is to make some assumptions, for example that people give answers that are lower than what they actually might be willing to pay, hoping that the eventual price would not be too high. Obvious management judgement and marketing research expertise is needed to interpret the early data for any given country and the market involved. Then an initial price (or in most cases a series of prices aimed at designated segments of course) is set.

Tariffing

The key is to start from the lowest end of the *acceptable* price scale and be prepared to adjust it upwards. Early on with mobile services, it is important to find early users to promote the service by word of mouth, to become evangelists for the service and get new users to the service. A low initial price will help the early adoption of the service, and create addiction with users. A low initial price will also deter competitors. They cannot base their business cases for competing services on anything but your prevailing price and being a follower, they must assume a significantly *lower* price for their competing offering. They do not know that you are prepared to make price adjustments upwards. The worst case is to price too high, frighten potential customers, and invite lots of competitors who can build profitable business cases by pricing below your price.

The vital key to introductory pricing is segmentation. The service proposition should be tailored to as precise a small segment as possible, with noticeable differences and distinctions to those services offered to other segments. This allows testing a variety of price ranges. Later several segment services can be combined into services for clusters, and prices adjusted at that time. If a price rise is combined with a service definition change — adding features for example — then the price change is likely to be accepted with little resistance. To find the optimal price and feature set, a series of services needs to be launched to a series of customers. Accurate segmentation is absolutely crucial to the success of tariffing.

14.3.4 Penny for your thoughts

With mobile services, the effects of optimized pricing are quite dramatic, partly because 1 penny is proportionally much more significant at price levels of 10–20 cents than the same penny is at the price levels of 5 dollars. Much more significantly in cellular telecoms, the potential customer numbers are quite dramatic, creating vast flows of money from tiny amounts.

A difference of 1 penny in pricing can have a huge effect on a popular service. For example, assume a new service is priced at 17 cents per use, and it is rapidly adopted and soon used by 10% of the subscriber base, who use it once per day. Also assume for the sake of simplicity that there is the ability to charge 18 cents for that service at no significant loss of use. Let us see how that affects a typical operator in a large European country like Italy, France or Britain. A typical network operator would have about 10 million subscribers, so 10% of that is 1 million users of the service. At usage rate of once per day, the service would be used 30 times per month or 30 million uses per month in the network. If the price was too low by 1 penny, the network operator — and any related content and application partners involved in a revenue share — would abandon 1 cent per user per day, or 300 000 dollars per month (30 million times 1 penny = 300 000 dollars), i.e. 3.6 million dollars per year of pure profit. Mobile telecoms turn pennies into millions quite literally, the user numbers are that dramatic. The tiny amounts, multiplied by the dramatic subscriber amounts, yield quite significant numbers in revenues. This same magic of micropayments also works in the profit and tariffing calculations. Every penny counts, very significantly, in mobile services.

14.3.5 Pricing of bundles ('service packages')

A particular early favourite of new service portfolios is the 'service bundle' sometimes called 'service package'. Bundling services makes bundle pricing an issue. It is too easy to collect a dozen services, bundle them together, and then say 'I'll give 10% off the whole set if you take the bundle'. Telecoms operators have been experimenting with this type of bundling pricing with long distance and friends-and-family, etc., types of early service bundles/packages. Bundle pricing is its own special science, but the very short discussion is this: make sure every component in the bundle is actually needed in the bundle, and that each service is separately analysed for network load and profitability. A good bundle attracts customers, binds them to the operator, and allows use of *new services* that the subscriber otherwise would not trial or use. A bad bundle gives extra discounts to customers who would use those services anyway.

14.4 Preparedness for tariffing

Tariffing is critical to profitability, yet its margins of error are very small. Price too high, and you lose customers, or the service will not take off. Price too low, and you give up profits. How can the 3G network operator solve this dilemma? The solution is a mixture of market research, tariff modelling, trials, and adaptation.

14.4.1 Marketing research

The first step in achieving a succesful price for any segment is market research. Any services that have a significant impact on the network should

Table 14.1 Typical service pricing — Iobox

Service items	Price UKP	Price USD
E-Mail forwarded to mobile	0.19	0.28
Send an SMS message	0.06	0.09
Calendar reminder via SMS	0.19	0.28
Send event by calendar	0.19	0.28
Send contact by SMS or card	0.19	0.28
Icon, tone or picture message	1.00	1.50
Mobile chat	0.06	0.09

(BWCS Mobile Matrix, 2002)

be measured through professional marketing research. Surveys and opinion polls can be conducted to test various service tariffing *types* — i.e. per usage price, monthly price, per minute price, sponsored service, etc.; and especially to test the tariffing *levels*. Tariffs should be tested not only in direct questions such as 'how much would you be willing to pay for this service' — but also through indirect means, such as testing the price level against existing service prices, such as SMS text messaging, etc. For any existing services by competitors, the competitor prices are of course also of great importance.

The network operators should build relationships with one or two major market research companies for building consistent surveys using the same procedures for the service portfolio. Results should be directly comparable across the service portfolio, and over time. Major content providers should find market research partners that can offer services in many of the content provider's significant target markets, in order to provide results that are comparable across international borders.

14.4.2 Tariff modelling

A very useful tool for evaluating tariff effects to profitability on a portfolio of telecoms services is tariff modelling. Many telecoms product-management departments have rudimentary simple tariff modelling tools, mostly simple spreadsheets to perform 'what if' calculations. More advanced tariff modelling tools are available from some of the specialist consultancies serving the telecoms industry. In designing a new portfolio of services, such as in introducing 3G services, the telecoms operator is prudent to use powerful tools to optimize its tariffing structure. Remember that a penny off of optimized pricing is easily millions of pure profit on an annual basis.

14.4.3 Tariff trials

When introducing new services, it is crucial that tariffs are announced as introductory and that tariffs are adjusted as user data comes in. For the goodwill especially of the 3G operator, who is perceived to be controlling the prices of all the services, it would be good if the 3G operator would have approximately as many prices going down as going up, after such trials. The operator should take the introductory pricing decisions as a learning process, and try to get the initial tariffs as close to right as possible, rather than every time giving it for free for 3 months and then introducing an astronomical price.

The time to use a 'try it for free for a month' or more approach, is when the service is genuinely new with no obvious near substitutes. Video calls on 3G networks could be such a service, as well as various multimedia messages, But if the service is only a minor adaptation of a similar and popular service on the fixed Internet, or in the newspapers, radio, etc., then there is no real reason to give it for free initially. The tariffing managers should be motivated to hit the right tariffs from the start, to invest good research, analysis and reasoning to find the optimal tariffs from the beginning.

14.4.4 Tariff adaptation

Finally, tariffs will need to be adapted to customer feedback and to market pressures. Here it is vital that the service creation system has a rapidly adapting structure to change prices. Many telecoms prices are so deeply ingrained into switching systems and monstrous charging and billing systems, that a price change literally takes months to implement. That is totally impossible for survival in the 3G services area. With five or six 3G operators and numerous MVNOs and other players introducing mobile services, reaction time has to be in days not months. Changes in tariffs have to be fast and nimble.

14.5 How about one price for all?

In every major telecoms market with open competition, there are players who are offering 'only one price', often suggesting that it is both simple and cheap for the user. When one examines such operators, it usually emerges that the 'one price' is only offered for a mainstream mass-market product (such as long-distance voice calls) but even these operators have special services priced differently. In most cases they are also willing to offer bulk discounts or other incentives for particular customers, such as large corporate customers, which means that the argument for one simple price is actually only a ploy to convince one part of the total customer base that this operator is the easiest to use.

'One price for all' can work for a player even in the mobile Internet services market, if that operator offers a simple portfolio of, mostly, mass-market products, does not compete with high innovation services, and is best suited for a player aiming for price leadership. However, we have to keep in mind that in the mobile Internet space, even for a limited portfolio of services, there will be numerous ones where strictly speaking one price may be impossible. For example m-commerce, if we have a single set price for 'all' activity on the mobile Internet, what if I access a vending machine to purchase a Pepsi? Is there still a separate charge for the can of the beverage or will all such small transactions be bundled into my one fee? When mobile services were mostly voice and SMS traffic, the argument for a simple price had a lot of merit, but how about sending picture messages from my phone to a friend who is on another network or listening to music which is streamed from some 'm-DJ' on a virtual m-radiostation in New York City? The scope of different services all working on different value systems is so broad that a simple 'we have only one price' may be a nice thought but very difficult to achieve without a lot of customer confusion and complaints.

14.5.1 Pricing by data traffic

A difficult proposition is pricing by megabyte. For business solutions and in situations when the buyer is a data-communications professional, a megabyte-based pricing model may be suitable, even preferable, but for mainstream mass-market products, the uncertainty behind megabyte pricing is very likely to introduce fear and concern in the price levels. Any such concern would diminish usage.

How to price data services? Numerous trials of data service pricing have been run in Europe, Asia and America. No definite preference has been established, but at least the idea is that pricing should be below a pain threshold and easy to understand. Tariffs should not introduce unnecessary concern, acting as an artificial barrier to service usage. One way is to show the price at related links on a service. As accessing any page costs on the mobile Internet, perhaps the link could show the price. As long as the individual pages are priced at rates ranging from fractions of a penny to a few cents, then it could guide users into accessing pages and sites. The big concern is that while the content prices tend to be better known, the data transmission prices would depend partially on network performance.

14.5.2 Home zones and hot spots

In many countries local calls are free or at very low cost. When they are low cost, a powerful tool to cannibalize fixed-network traffic to the mobile network is 'home zone' tariffing. Home zones are not technically particularly complex, they tend to be purely billing solutions where the mobile phone is charged equivalent rates to the local call when calls are placed within a specific geographical location such as a person's home. Home-zone tariffing will need to be calculated to remain profitable, if only marginally so. The reason is that the user is taught to abandon the fixed line and only use the mobile phone. In the home-zone area, like the home or office, the user will have no reason to use the fixed phone, as the mobile phone rates are the same, but the mobile phone provides considerable utlity over the fixed phone. The mobile phone then, of course, becomes the natural phone to use when stepping out to the store, etc. The moment the user steps out of the home zone, regular mobile tariffs apply. Depending on how the balance of the traffic is distributed, this customer can be slightly profitable or reasonably profitable, but never as profitable as a mobile customer who does not have a home-zone service. The home-zone pricing strategy is considered as a defensive move against fixed operators or operators with fixed and mobile services.

With Wi-Fi (wireless fidelity), also known as W-LAN (wireless local area networks) or by its standard 802.11, it is possible to deliver high-speed data in small areas like a café, hotel lobby or an airport lounge. Early Wi-Fi business models give near unlimited data loading at a small fee, for example the cost of a cup of coffee. These localized hotspots as centres of high telecoms traffic loads can pose a problem for mobile operators. To compete with hotspots, similar pricing strategies can be used. Even with 3G/UMTS and using microcells and picocells in the cellular grid, a 3G network could not achieve similar throughput to a Wi-Fi hotspot. However, with 3G phones soon numbering in the several hundreds of millions, and Wi-Fi devices unlikely to even reach 100 million by 2006, a much larger installed population is accessable via 3G/UMTS. The operator could offer reduced rate pricing near known public hotspots such as the airport, major hotels, etc. The mobile operator would then include some added features — such as offering the hotspot at twice the area offered on Wi-Fi. This could be for example that if the Wi-Fi is limited to the lounge area of the airport, the 3G hotspot could start already at the car parking lot.

The 3G operator would trade service speed for mobility — the natural strength of the cellular network — deliver value to its customers, and take some customers away from the Wi-Fi operator. If the price and performance differential is very large, it would justify purchasing laptop computers with Wi-Fi connectivity, or getting Wi-Fi cards to laptops, but if the price and performance of the 3G network is not perceived to be much different from Wi-Fi in the hotspot areas, then an infrared or Bluetooth connection for the laptop to the 3G phone would be the common solution for most.

14.6 3G licences and the price

The 3G licence wars and of course the new infrastructure investments in countries like the UK and Germany make it impossible to set the 'actual cost' price tag accurately onto the first services to come, but the calculation has to be extended to longer time periods. It is not a matter of 3G not being a profitable business in these countries, but that it will take longer than expected in order to be so. The first and the coming up services, too, will therefore certainly be sold at what is initially likely to be unprofitable prices. This is an idea that we just have to face and get used to it. Of course most of the world's countries did not have particularly large 3G license fees, so this concern does not apply to the vast majority of 3G operators.

14.6.1 Price 3G for mass market adoption

In tariffing mobile services it is vital to keep in mind that mobile Internet services for the most part are mass-market services. Their true power is unveiled when hundreds of thousands, or preferably millions, of people use them frequently. The pricing will need to be set very low so as not to hinder usage and to allow word-of-mouth and viral marketing to take care of publicizing the new services. To summarize this chapter, we could say use tariffing that is just below the 'pain threshold'. The clear example of how powerful this can be is SMS text messaging. As text messages typically cost between 10–15 cents per message, they are comfortably below what almost any person would consider to be a prohibitive cost per usage. Of course, when the phone bill arrives and it has 50 dollars of SMS text messages, we may stop and think that we should cut down, but at 10–15 cents per message, the individual cost is simply so small that we don't bother to worry about it.

That is how mobile services should be tariffed. Keep the price below the pain threshold. If not sure, aim *lower* and then build the traffic and adjust later. Don't stifle the usage by tariffing too high. A good example of how wrong operators got their pricing was WAP pricing. Here users were waiting to receive some information from a service which seemed extremely slow, and became increasingly frustrated and even angry when considering that every moment of delay was costing the users.

14.6.2 Not a free for all

This chapter has looked at tariffing. We looked at how the airline industry has managed to convince most passengers that their pricing is simple and straightforward, while being incredibly complex with literally thousands of prices on any given route. Telecoms operators must learn to use tariffing as a tool for profitability and learn to use segmentation, bundling, introductory offers, etc., to create the right mix of prices for services. Tariffing has a bigger impact to revenues and profits than any network optimization or fine-tuning of the marketing budget. And prices lead directly to money coming into the company. Lets keep in mind what Woody Allen said: 'Money is better than poverty, if only for financial reasons'.

Sales executives should be optimists, and credit controllers should be pessimists.
— Roger Regan

15

Billing
Don't Forget Billing

The billing and charging ability of modern mobile telecoms operators is in its own class. Compared with any other recurring billing by national companies that touches all of us, telecoms operators collect, itemize and bill the smallest units of service and are able to report on the transactions to a remarkable amount of detail. No other industry tracks and bills transactions of such small value such as is done in telecoms. Charging and billing are true competitive advantages for mobile operators yet billing and charging are not

---

3G Marketing: Communities and Strategic Partnerships Tomi T. Ahonen, Timo Kasper and Sara Melkko
© 2004 John Wiley & Sons, Ltd ISBN: 0-470-85100-7

well understood and often seem an afterthought to service creation. Many real examples exist of new telecoms services having been introduced that could not be invoiced, at least not initially.

15.1 Charging, billing, reporting

Charging, billing and reporting are often used interchangeably and some may think they are synonyms. They are not. At the heart of telecoms billing lies the charging engine. The charging engine collects the data. From the charging information, a bill is prepared. That is billing. Billing creates a payment obligation on the subscriber, and if a summary or further itemized details are requested, then that is reporting. Reporting replicates information that was on a bill, therefore there is no payment expected of reporting.

15.1.1 Charging collects the data

Charging is the activity of collecting proof of billable actions, and in telecoms charging generates CDRs (charge detail records also often called call detail records). The CDR contains all necessary information for one chargeable item, i.e. one phone call or one SMS text message. It would state which number placed the call, which number received the call, at what time the call started and ended, what tariffs were in effect, etc. A CDR is generated for every call, resulting in masses of data entries for items of miniscule value. A typical cellular phone call can be 30 seconds in length, go to another person in the same network, and in a very real sense, be worth only pennies.

It is important to note that no other industry tracks its product sales or service usage to the value of items worth pennies. The paper-clip industry does not track by paper clip, but rather by boxes of 1000 paper clips. Your electricity is measured in kilowatts, not watts. The banking and credit card industries would potentially need to track items worth a penny but few people do banking in denominations of a penny, and credit card companies impose a minimum payment on their charges. When you consider that a typical mobile phone user can place a number of calls per day, and any one of those calls can legitimately be worth much less than a dollar, you can understand that the telecom operators have a powerful asset on their hands. The ability effortlessly to handle tiny payments produces a competitive advantage in handling so-called 'micropayments' or payments worth less than 1 dollar. Micropayments will be discussed later in this chapter.

15.1.2 Billing creates the invoice

After a CDR is generated, a bill can be prepared. Billing is done in cycles such as on a monthly basis, and all relevant charges are tallied for that period and collected into a bill or invoice. Billing includes adjustments such as various negotiated discounts, any corrections that may relate to over- and underpayments in the past, etc. Billing as such is a relatively simple addition and printing operation if the charging is done correctly. While people often talk about telecoms-operator billing being a near-draconian system, it is actually the charging engine which is of incredible scale. The billing ability to generate a few million invoices every month is nothing that other large national companies serving consumers could not do.

15.1.3 Reporting gives information to the caller

Reporting is the activity of summarizing data that relates to a bill. It may be in conjunction with a phone bill, usually called itemized billing, or it may be separate, such as calling reports that are generated for corporate customers to track their use and costs of telecoms, or for individual users who may want a year-end summary of the telecoms spend. Reporting has become an important value-add service in Japan with Mobile Internet services, as consumers are making m-commerce purchases and are using various new services on their mobile phones. As more services become possible on the mobile phone, it also becomes more important to be able to see what all was consumed in any given period.

15.2 Micropayments

The biggest competitive advantage that the mobile operator has over all other digital content delivery systems is not location information, nor the personalization information; it is micropayment billing. The telecoms operators are the only entities that currently control a billing system powerful enough to handle micropayment amounts — charging/billing records with a value of less than 1 dollar — on a system that scales to *millions of users* and easily dozens, even hundreds, of micropayment record entries per user *per day*.

The micropayment engine of the billing system can provide numerous solutions for billing for services and content that has a real value only measured in pennies. This could be, for example, billing by web page viewed.

No other billing system, not even those used by credit card companies and major banks, is robust enough to handle transactions of such small value.

15.2.1 Credit risk

When taking on activities like those of banks and credit card companies in handling the payments of mobile commerce purchases, the mobile operator (or portal) is also entering into the area of credit risk management. This is the natural domain of banks and credit card companies. While mobile operators have also been involved in the credit worthiness of their customers, mobile purchases would introduce added risks. The change would in many cases include requirements of registration with the banking regulators. Some of the particular issues would involve large-value credit card purchases, which may well be in excess of the typical values of telecoms transactions — such as purchasing airline tickets.

Other concerns may be with prepaid accounts, for example, is the real time prepaid account counter able to recognize exactly when the account has reached zero and not allow high-value purchases. Many prepaid accounts will allow the last call to be made when the account still has some pennies left in value. Those pennies cannot be used to pay for a purchase valued in dollars. A further concern with prepaid accounts is that of any discount given to top-up voucher purchases. If vouchers can be purchased on discount,

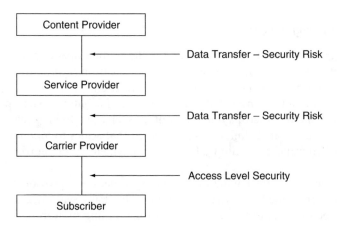

Figure 15.1 Security risks in the business chain

then an arbitrage situation might arise where a user buys a voucher for 5% discount, then goes to a cash machine, withdraws the full account making 5% profit, then goes to voucher dealer and buys another voucher, and again returns to the cash machine, making 'free' money at the expense of the mobile operator.

15.2.2 To bank or not to bank?

A separate idea is whether or not the mobile operator, MVNO (mobile virtual network operator) or portal wants to offer banking and credit services under its own brand. In most markets this requires a banking licence but it is not that simple. Banking in itself is not profitable in most markets at the level of moving small amounts of money. Banking profits come from the handling of credit and other financial instruments. So for a mobile operator to make money out of being a bank, it is not enough to have an efficient billing system. In fact, the critical part is to hire competent credit and financial staff and to set up efficient modern banking processes. One could argue that, for example, the consumer credit card business is so tightly contested between the big global players like Visa, Mastercard, Diners Club and American Express, that any newcomer like a mobile operator is likely to lose very much money trying to get into the business. Inevitably the customers most attracted to the newest credit player are of course those who cannot get credit with the existing players, i.e. very bad credit risks. This book is not about mobile commerce or mobile banking, but many who think about billing assume it is a quick way into that business. We suggest that these think very hard and examine the competitiveness of regular banking and the credit industries before taking major steps in that direction.

15.2.3 Tracking advertising and promotion revenues

The billing system with most mobile operators will need to evolve further to enable the full functionality of 3G services. The advent of sponsored content and services brings new challenges, such as McDonald's restaurants sponsoring the local city guide map so that the local map is free for users but displays all McDonald's restaurants on the maps, of course. This way the service usage needs to be tracked and billed to a third party — McDonald's this month, perhaps Pizza Hut next month. Similarly, when customers view advertising, it may generate points or even money that needs to be accredited to the customer. Some ads may feature coupons that need to be tracked, timed, used up and terminated when they are redeemed. Any viral advertising — where a user may forward a coupon to several friends and each recepient may use the coupon at full value — and better yet, where the customer forwarding the ad would be rewarded as well — will need to be tracked and correctly credited. The third-party revenues provide a new area which in most cases was not relevant to telecoms in the past.

15.2.4 Tracking digital rights

The digital rights management (DRM) interests of content owners and any digital delivery media comprise another area where mobile operators need to evolve. The billing system is a natural element to keep track of what content was delivered to which user, whether it was forwarded further and if so, to ensure that the appropriate charges are levied also from the new user. DRM is a new area in which the digital content owners approach the mobile Internet as the most secure networked delivery mechanism, but for the mobile Internet to deliver, the charging engines of the mobile operators will need to be upgraded to track digital rights.

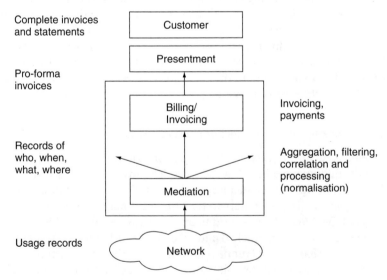

Figure 15.2 Billing System for Portal Services

Source: UMTS Forum Report # 16, November 2001

15.2.5 Billing can also be an added value service

While often considered to be the least exciting part of telecoms, billing can be a service by itself. The charging information, either in the bill or in separate reports, can be used to provide valuable information to users, especially for business customers. As ever more of the workforce's services can be purchases via the mobile phone, then the collecting and reporting ability of the charging engine can be used to deliver details by user, by department, by customer account code, etc. Such details can provide significant accounting information for business customers, and services built around them can be billable, profitable and binding services.

15.3 From billing to product management and marketing

Charging and billing data provides crucial customer information for development of the operator's service portfolio. The charging information can be used to estimate market potential of different services and can identify opportunities future for products that are likely to be successful. Therefore, service management and marketing should have access to this data or at least get valuable statistics regularly.

Ideally, the customer billing data is linked to a CRM (customer relationship management) system. In the the identification and definition of product needs billing can provide critical information for segmenting the customers. For better and more focused marketing activities in general and sales activities in particular, billing and CRM can give hints for cross-selling to existing customers and in finding new target markets for existing products. For example, customer segment X, teenage girls, is a heavy consumer of the service A, mobile chat. Another segment, Y, young female professionals shows the same characteristics as segment X, but also uses the value added product B, cartoon character picture messaging. This gives the opportunity to offer product B to segment X, the teenage girls. Such analysis of customer segment behaviour can identify specific killer applications for segments that have similar needs.

The information contained in telecoms-operator charging systems is a marketing goldmine that will be 'mined' both by using data-mining tools and more advanced segmentation systems such as self organizing maps (SOMs). The patterns and opportunities that can be identified will enable product and service development and management to offer services ever more accurately to customers that want them and are willing to pay for them. It all

Table 15.1 Revenue composition comparison

	USA	Europe	Japan
Penetration	50%	80%	65%
ARPU (Average revenue per user)	$52	$33	$60
Messaging	1%	13%	7%
Other mobile data	1%	2%	13%
Total mobile data	2%	15%	20%

Financial Times 28 October 2003

comes back to customer intelligence and segmentation as we discussed at the beginning of the book.

15.4 The call for one bill

Customers increasingly demand consolidation of their data transfer phone bills. In most markets customers still get different bills for their fixed-line telephone, their mobile phone and their Internet usage. This is often so even when all of these services are provided by the same company. Customers should not be forced to handle separate bills from the same company. The consolidation of billing can be a billing system headache but by now most telecoms operators must have the technical ability to provide a consolidated invoice to all customers.

The sad truth is that often in companies where separate units operate, and bill for, fixed telecoms, mobile telecoms and fixed Internet services, those units are also oblivious to the true customer nature of any given customer. Where that customer might bring very small amounts of fixed telecoms (voice) traffic and by that volume would be classified as a non-important customer, the same customer might be a heavy user of Internet services or mobile voice, or both. That customer will perceive all service from any unit of one telecoms operator as the behaviour of the corporation. Bad service by one unit can do irreparable damage to a high volume, high value customer of another unit or units.

Technically, from the charging and billing point of view, the same product can have different prices and this should be made possible in the billing. With mobile voice, home-zone type pricing plans are typically built around this premise. The service is identical in any cell of the operator's network, to a voice call, but by designating one cell or even more precise location information as the home zone, that area is given less expensive voice minutes than calls placed elsewhere on the network. Technically, the voice call is the same anywhere but by manipulating the tariffing records lower costs can be delivered in one (or in a designated number of) cell(s). The more attractive pricing model is primarily a marketing model for tariffing, with little if any 'real technical merit'. This may be a difficult concept for traditional engineering mindsets to get to terms with, but as is true of all industries, marketing plays a bigger role in the customer's perception of a service or product than strict performance data. Thus, with the history of geographical pricing in telecoms such as paying more for long distance and even more for international calls, customers are very willing to accept the premise that calls in one region, location or place are priced differently from those in another place. This ignorance in the minds of consumers should be exploited by the cellular telecoms operators and services like home zone be deployed.

15.5 Revenue assurance

A major concern with telecoms operators is revenue leakage. While fraud is often considered to be synonymous with revenue leakage, actually revenue assurance is a much broader issue than mere fraud control.

Global acountancies such as PriceWaterhouseCoopers have estimated that cellular operators have revenue leakage at a rate of about 5%. This means that services are not billed at all, are billed the wrong rate, are assigned the wrong discount, or the payment is not received. Some customers typically are charged too little, while others are charged too much. Both provide problems to the telecoms operator. Charging too little means lost profits. Charging too much exposes the operator to the risk of challenges to the billing and even lawsuits. Revenue assurance is the process of identifying the various leakages and plugging them.

It is important to understand that with revenue leakage, the customer has consumed services and is perfectly willing to pay for the usage. It is the network operator's inefficiency that results in consumed services not being billed correctly, with the correct tariff and potential discounts, for the full duration of the usage. As the CDR generation process itself is — or should be — totally automated, then any error becomes automated as well, meaning that the effect of errors in the system is multiplied. While in most cases revenue assurance involves stopping leakages — of services not billed — errors in the billing system can also generate opposite problems. In some cases and countries, if the phone bill is too large, it may even set the operator subject to a potential lawsuit.

15.5.1 Revenue leakage and profit

One cannot stress enough the direct relationship between revenue leakage and profit. The revenues that are not correctly billed, and thus are forever lost, are *profits that are abandoned*. The effect of discovering revenue leaks has a direct effect on the bottom line, as few other activities in telecoms can possibly have. If we assume that a mobile operator has 1 billion dollars of annual revenues, and its leakage is 5%, then the operator abandons 50 million dollars per year.

The revenue leakage effect is even more pronounced if we look at profitability. If we assume, in the above example, that the operator had a 20% profitability, generating profits of 200 million dollars, then the revenue leakage — 50 million dollars — cost the operator one-fifth of its profits. To put it in another way, just by plugging the leaks, the operator's profitability would leap from 20% to 25%.

While evaluating the accuracy of telecoms charging and billing systems requires a special skill, with new technologies it is becoming increasingly possible to evaluate the full system. As the effects an the bottom line are so dramatic, operators are encouraged to identify and plug their revenue leaks now.

It is important to understand that some sophisticated business customers are on the way to doing that for themselves. With various systems tied to their IT and telecoms structures, some sophisticated corporate customers are able to track telecoms usage and compare it to the invoices. In the future and with services built with IP, this will become ever easier for the customer.

15.5.2 Revenue assurance and 3G

Revenue assurance becomes even more of an important issue with 3G services as the whole application and content provider dimension enters into the telecoms traffic and billing picture. Before in 2G, the network operator controlled the full value chain and billed — or forgot to bill — for all services. With 2.5G and especially 3G, there will be application developers and content providers who expect to be paid. The revenue sharing arrangements can very easily be quite removed from actual billing data. Any errors

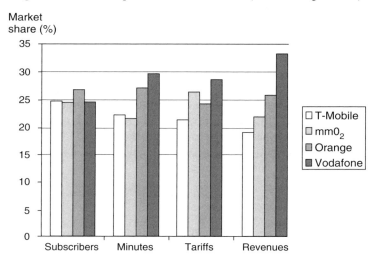

Figure 15.3 UK comparison of market shares (OFTEL August 2003)

in the charging and billing data are magnified once again, where the original customer is not billed for services consumed, but the content partner is paid for them. The operator can be hit twice with the error.

The advent of m-commerce and mobile advertising will add further complications to the billing situation, creating massive additional opportunities for lost profits due to leakage. With m-commerce, the money transactions may be large while the relative proportion of the operator's share is small. In that case even a very small error can eliminate all revenues generated by any given service. With m-banking and other mobile finance services, the issue is even more critical, as the data transferred may be trivial, the payment collected by the network operator equally a mere token, but the value of the reported transaction can be quite considerable. Errors in m-banking can threaten the profitability of all of m-commerce.

15.5.3 Billing complaints

With reference to revenue leakage and inaccurate billing, the process of how to handle complaints should not to be neglected. A complaining customer does not threaten to leave a network just because they has an issue with the bill or any other matter. It is very important however, that the issue is dealt with promptly and to the customer's satisfaction, otherwise even one issue can grow into a threat to jeopardize the customer relationship. The chapter on churn (Chapter 17) deals with how to handle customer complaints in more detail. Separately, there are complaints that are not billing related. Those should all be carefully collected and forwarded to service development teams for consideration on service evolution as we discussed earlier.

As telecoms billing systems are remarkably complex, the question arises as to whether or not customers can understand their phone bill, or determine if the bill is accurate. Operators should not hide behind complex and incomprehensible phone bills. Even though the charge detail generation system is incredibly complex, the final phone bill presented to the customer should be very simple and easy to read.

15.6 End to billing

Charging, billing and reporting tend to be considered 'unsexy' in telecoms and given little thought in service creation. The systems themselves are enormous and changes typically involve man-months of work and large costs. The charging and billing system is a unique competitive advantage that telecoms operators have over all other players in 3G. This advantage should be capitalized. The reporting system can also be used in some cases to generate added value to customers, content or application partners.

Furthermore, careful attention should be given to revenue assurance to plug leaks directly from the profits of the company. Charging, billing and reporting is not glamorous but it is, of course vital for collecting the money for the business. The database generated by the telecoms charging system is the envy of all other industries and a true gold mine for the industry. However, statistics can be interpreted in many ways, as comedian Jay Leno said: 'The *New England Journal of Medicine* reports that 9 out of 10 doctors agree that 1 out of 10 doctors is an idiot'.

It is not fair that only one company makes the game 'Monopoly'
— Steven Wright

16

Other Revenue Streams
Partnerships

In most cases of services for 3G the primary revenues will come from the end-user or whoever pays the phone bill, such as the employer or parent, etc., The majority of 3G revenues will thus be paid as subscription fees, usage fees, etc. In 3G and more widely in the mobile services space, significant new revenue streams will emerge from payments from sponsors and mobile advertising, sales commissions from mobile commerce transactions, etc. This chapter will take a look at those emerging new revenue streams in 3G and the inevitable partnerships that will be needed. This chapter also discusses revenue sharing that is involved in such partnerships.

16.1 Redefining the operator position

The network operator's role used to be very clear in the past. They used to grow in a 'queen bee'-like scenario — big, protected markets with very little or no competition, at least with a limited and well defined number of other operators as competitors. The more or less oligopolistic market pampered the network operators with massive profits and therefore the philosophy of sharing revenues does not come easily to mobile operators. However, as we have stated at many points in this book, 3G will not be a commercial success without the cooperation of a large number of different players all along the value chain(s). Operators and their ability to create services on their own simply are not enough for customers.

Fixed and mobile network operators used to gain revenues mainly from monthly phone-line charges, calling costs and, more recently, data transfer. Internet service providers (ISPs) then emerged on the markets to offer Internet access. Soon they expanded their market opportunities by leasing excess network capacities from network operators and offering access services, telephony services and content. The entry of ISPs altered the revenue equation. The network operators faced the following trade-off: give up a part of their existing customers to the new ISPs versus collecting revenues from unused parts of the network. The new revenues could be used to pay back the costs of upgrades to the telecoms network and thus partly recover infrastructure investments.

In the 1990s with the emergence of ISPs, the risks to fixed network operators seemed rather limited. The new ISP's target customers were not exactly a subset of the fixed network operator's existing client base, but might attract totally new target markets. As competition emerged in fixed telecoms, the competition affected entered the equation: the *other* competing operators might lose *even more* in market share. Even though ISP's typically sold services at substantially lower prices than the incumbent operators, the fixed operators could still count on customer retention serving their competitive position. Still today the bond between operators and their customers is very strong as far as, for example, access and data services or billing is concerned.

Figure 16.1 European mobile content in 2003, worth $1.1 billion (Jupiter Research March 2003)

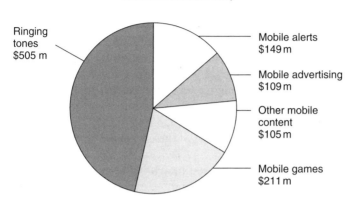

16.2 Business models

The second generation of mobile telecoms and their digital services introduced revenue-sharing models through trial and error that now are evolving. These will be transferred to the 3G world. Sonera ZED and Jippii, who were global pioneers of new mobile services in Finland, can act as good examples of how similar services — offered alternatively by the operator or a content provider — can generate significant revenue streams. I-Mode in Japan can show an alternate way to approach the market opportunity.

16.2.1 Case Jippii Group

As we explained in the Introduction, the whole mobile services industry was created by a fixed Internet service provider in Finland now known as Jippii Group. Part of the innovation was in their new revenue model. Jippii, or what was then called Saunalahden Serveri, started out as a merger of several small ISPs in Finland. As they initially targeted only the domestic consumer, the internationally challenging name, Saunalahden Serveri, was selected to convey a very domestic Finnish image. The name literally means 'the server sitting by the Sauna-bath at the bay'. The merged company was soon the third largest ISP in Finland and became known for its youthful image and innovative services. Its primary rivals were the ISP arms of the two biggest incumbent telecoms giants in Finland: Sonera and Kolumbus of what is now known as the Elisa Group. Both Sonera and the Elisa

Group were full-service telecoms operators with local, long-distance and international calls, mobile networks and a broad range of datacoms services. Jippii Group was only an ISP.

Jippii experimented with various telecoms services in Finland including getting into the then-lucrative international calls business. Then, in 1998, Jippii entered the mobile telecommunications business in what was seen as a totally marginal niche market and at best a short term fad — offering downloadable logos and ringing tones to selected mobile phones, transmited via SMS. This business was dismissed by most analysts, as in 1998 there were only a handful of phone handsets where changing the ringing tone was even possible — and most of these were the most expensive top-end models. The same was true of the lack of phones accepting changeable logos. This was not seen as a good match between 'youth' customers who might want to buy music or logos, and the expensive handsets necessary.

Jippii Group was not deterred. They were operating on the economics of the ISP business, not mobile telecoms operators. Where an operator would want millions or at least 100 000s of customers for any given mass-market product or service, the ISP could be very happy with 10 000 customers or even fewer. Silently and almost in secret, Jippii started to capture the hearts of the youth in Finland. Ringing tones started to emerge with chart topping hits of the biggest pop bands. By 1998 already over half of Finns had mobile phones and the 'me-too' effect of wanting ringing tones that sounded cool was spreading fast among the youth segment.

After a few months of Jippii having a monopoly on the services, the big players, Radiolinja (owned by Elisa) and Sonera both introduced ringing tones and logos. The service usage then ballooned and a real, unexpected killer application had emerged.

The business idea behind these services was to offer SMS based services to the mobile operators' customers by using the mobile operator's network and technical solutions, i.e. through its smart message service center (SMSC). Via leasing agreements, the content provider — Jippii — could offer SMS services to customers by using so-called microbilling: the mobile operator bills its own customers in the monthly invoice, for example 'third-party SMS services', and the revenue is shared between the parties.

Jippii Group found the business model so promising that the company changed its name and started to introduce ringing tones, logos, and other advanced services in other countries in Europe.

In the original arrangements in Finland the revenue-share received by Jippii — and soon many other service providers as well — was small. The mobile operators were in a strong position when at that time in Finland there were, for all practical purposes, only two mobile networks and the operators took a very protectionist position in revenue sharing arrangements. However, as ISPs were struggling to generate any revenues — this was the time when the free model of Internet access was at its peak — any revenues were better than nothing. ISPs were generating what for them were very significant revenues through the revenue-sharing agreements of mobile services on SMS. A key achievement by Jippii's innovation in Finland was to generate a mindset of sharing with mobile services.

The power of the consumer demand was soon noted by Nokia with its headquarters in Finland. Nokia was quick to spot the opportunities of ringing tones and logos and to capitalise on them. Very soon most Nokia phones, including all targeted at youth segments, could accept ringing tones and logos. As word of the success of Jippii Group and the bigger players making money on advanced SMS services in Finland spread across the world, so too soon the ringing tone and logo success followed. The next year saw the introduction of I-Mode by NTT DoCoMo in Japan and it is no surprise that among the most popular services on I-Mode — and its rivals J-Phone and KDDI — today are still ringing tones and logos.

16.2.2 Case Sonera Zed

As the Jippii Group was nimble and quick to make money on niche opportunities, Sonera launched its Zed unit to deploy mobile services. In fact, most operators in early-adopter countries found out the potential of these SMS services and started to form their own service portfolio or even mobile services brands. In Finland, in addition to Sonera's Zed, its rival Radiolinja of Elisa Group introduced its service under the brand of DJ Esko. In Sweden the Scandinavian giant operator Telia launched Speedy Tomato, etc. Sonera's Zed was particularly successful early on in Finland and rapidly expanded to pursue opportunities beyond Finland.

Zed's business model was to offer an SMS services platform to operators by creating services and a strong brand. The idea was that SMS service providers would then make contracts with the mobile operators. While nobody doubted the appeal of mobile services, the value-add of a Finnish mobile services brand was not seen to be clear. The expansion of Zed to other markets proved to be considerably more difficult than initially anticipated as, in most markets, there was no philosophy of open networks and third parties accessing the operator's customers. Thus Zed and many of its

Figure 16.2 Growth of official websites on I-Mode in Japan Official websites are the official approved content providers on I-Mode. There are also 65 000 unofficial websites providing further content for I-Mode

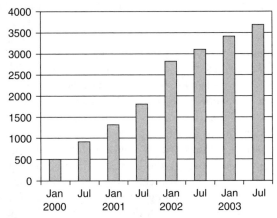

Source: NTT DoCoMo 2003

contemporaries had to resort to offering SMS services via service numbers in other markets than their own.

Even though mobile services were rarely successful beyond the home markets of the launching companies, SMS services were very important to the industry as they introduced the idea of sharing revenues and proved that both the network provider and the service provider could profit from such arrangements. SMS services were the first to use a revenue sharing model, which showed that the more competition and diversity of services the market offered, the more it grew and all parties could profit. Still, as the SMS service porfolio was limited and European network operators took very protectionist and 'greedy' revenue-sharing positions, true mass-market success was not seen from these early examples. The mass-market success was about to happen on the other side of the globe.

16.2.3 Case I-Mode

The first true mass-market commercial success for advanced mobile services was, of course, NTT DoCoMo's I-Mode service. With numerous innovations, NTT DoCoMo I-Mode's greatest contribution to the success of advanced mobile services was its radical revenue-sharing scheme. With I-Mode, NTT DoCoMo kept only 9% of the content revenues, returning 91% to the content owners. NTT DoCoMo was not being overly generous in doing so, as it still gets to keep all of the traffic revenues — the value of the data traffic on some services can exceed the value of the content — but for the content owners, for the first time, a mobile operator was offering a reasonable revenue-sharing proposition for established services available on multiple delivery channels. By giving away only 9% the content owners could easily survive and build better services. Soon the service portfolio on I-Mode was vastly beyond that on any other mobile network. With more services came more users, and with a larger user base, even more service providers were attracted to I-Mode.

The revenue-sharing model by I-Mode actively attracted new content providers. Major global brands such as Disney are prominent on the I-Mode service. As long as 2 years after launch, I-Mode was still reporting a delay of several months for new content providers to be connected as official I-Mode websites, so strong was the demand to join the community. At the time of writing of this book, the I-Mode content provider community was still by far the world's largest in the range of mobile content and services offered.

Of course the revenue-sharing model was not the only reason why NTT DoCoMo was so successful with I-Mode. Many other factors assisted the company and the service, such as the technical standard — I-Mode used standard HTML rather than the new language WML used by WAP. The Japanese market was unusual in its very low Internet penetration among advanced economies. With no SMS culture and little Internet access, I-Mode's e-mail service in particular took off much more strongly than did e-mail services on most other mobile services. Competitive reasons and the fact that I-Mode had custom handsets also contributed to its success, but regardless of this, the biggest single factor was that the revenue model attracted good content.

16.3 Operator revenue strategies

There are several different opportunities for mobile operators to make money out of 3G and advanced mobile services. Since it has already been stated that operators simply do not have the necessary resources to compete at every stage of the value chain, a thorough analysis should be executed to identify best opportunities. There are different alternatives that should be taken into account when analysing the strengths and unique service propositions of the company.

Table 16.1 Selected mobile virtual network operators from other industries

Cosmo Girl	Magazine, UK
Financial Times	Newspaper, UK
Kiss	Heavy rock band, USA
MTV	TV channel, Sweden
Narvesan	Kiosk chain, Norway
Nelly	Rap artist, USA
Stockmann's	Department store, Finland
Super Stable	Game, Hong Kong
Twins	Pop artist, Hong Kong

The most vital decision is about how much of the service creation is done in-house versus outsourcing via partnerships. The mode of operation naturally has a strong effect on the quality (and even quantity) of revenue. There is no general recipe that can be applied for all operators, but the different market situations, business strategies and know-how have to be taken into consideration.

At one extreme, operators can profit from their own brands and by creating exclusive services. In this way they can try to hinder third-party content providers from entering the market. Even though this might seem very appealing at first sight, it is almost certain that this business model will not succeed in the long run, at least with major categories of services. At the other extreme, operators can profit from acting as mere infrastructure providers and lease network capacities to third parties. However, this would mean reducing costs and effort but also neglect a substantial amount of more lucrative revenue streams.

Of course in most cases the mobile operators will find a middle ground. This means engaging in strategic partnerships and joint ventures and milking the revenues in selected stages of the value chain. Alas, the more content providers are given access to the operator's network and its customers, the more complex the revenue sharing models and coordinative work will be, but whether operators want it or not, the trend followed even by regulators is going towards liberalization and opening up the networks for MVNOs (mobile virtual network operators) anyway. With the 3G market constantly growing bigger and more complex there might be a niche for companies coordinating the various application provider agreements and taking over this work for the operators.

Operators have to see application providers in a totally new light, and not as competitors but as partners and even customers. Table 16.2 sums up the different revenue opportunities for operators:

We will look into some of the relevant items in more detail in the following.

16.3.1 Selling location data

The location business will clearly be a growing revenue stream in the future. The selling of location data works in two ways: first the operator can sell information to third parties about the user's whereabouts. Second, the user himself wants information about location. This can be in three primary areas:

Types of location services

(i) *Pull services*: where is the nearest hair dresser, which hair dresser around offers me a special discount?

(ii) *mapping and navigating services*: where am I if lost in a new city, how to find a friend's home, what distance have I walked, what is my travelling speed?

(iii) *communication services*: are some friends around here somewhere, where are my employees at this moment, dating services, communities, etc.

16.3.2 Location based push services

By selling location data to third parties, these can then offer services that might be intereresting to the user according to his location. A typical example is as you walk down the street you might get an ad of a nearby shoe store to say that they have certain types of shoes on sale right now. These push services, however, can quickly become annoying. Again, the line between information and advertising is very thin. Location data could be linked to promotion: a fast-food chain could subsidize the cost of some service on the user's behalf who, in return, agrees to accept a certain amount of advertising whenever he is around a subsidiary of that fast food restaurant.

Even more important than push service revenues, positioning services will be the next service revolution. The Global Positioning System (GPS) was developed by the US government for military purposes and at first it held a monopolistic position when it came to positioning services. That closed system had hindered service creation and widespread adoption by the public, not least because of rather high terminal prices. With 2G and 2.5G

Table 16.2 Summing up of revenue opportunities for operators

Operator's asset	Direct revenue and revenue sharing opportunities for operators	Partners
Infrastructure (network, transport)	Traffic revenues Lease fees for excess network capacities	Virtual mobile operators Application (service and content) providers
Billing	m-Commerce Banking and credit	m-Commerce companies Banks Credit card companies
Customer data (personalization)	Traffic Services m-Commerce	Application providers m-Commerce companies
Location data	Advertisement fees (push services) Selling location data (pull services), e.g. nearest restaurant, etc., intelligent information services	Any company (advertisers) Advertising agencies Yellow Pages Mapping services
Portal	m-Commerce Services Advertisement	Portal partners Application providers Advertisers
Applications/services	Traffic Service fees Billing services	Application providers

cellular networks, simple and relatively inaccurate positioning became possible, usually to an accuracy of about 50–100 metres (150–300 feet). This level of accuracy was enough to identify a major landmark for a pedestrian, but not enough to provide real-time driving assistance to a car. With advances in cellular technology the accuracy is being refined and in 3G the accuracy can be close to that of GPS. It will be very interesting to see how 3G will affect this business — whether in forming partnerships with GPS providers (terminals, maps, content) or by providing a totally independent alternative and competitive business or something in between.

With the growing mobility of users, positioning services gain momentum. After at first being used by special interest groups and business users, GPS has now started to explode and 3G operators have to think about their positioning strategies in order to gain market share. 3G operators will find a huge, fast-growing market place with positioning services that give them a key opportunity in handling the most important piece of information: namely where the user is. Strategic positioning could be executed with different kinds of partners: automobile industry, map providers, guiding services, employee monitoring, and security businesses. People with health issues could be given special care when vital to position them in time.

16.3.3 m-Commerce

Another area is mobile commerce. This is a whole area of its own and many books have been written about it recently. We simply summarize here that mobile commerce brings with it the opportunity for a network operator to get a small part of the revenue stream from handling the billing or even beyond. While many transactions are likely to be small in value, such as buying a ticket for the underground trains or paying for parking, the volume of transactions can be huge. Already, for example, in Finland with a population of 5 million, about 225 000 Finns play the lottery via the mobile phone. The revenue stream per user is tiny, but the aggregation of such streams brings significant revenues from m-commerce.

16.3.4 mAd (mobile advertising)

Mobile advertising, promotion and marketing is seen as a an area of great growth and opportunity. Some direct marketing activities can migrate to the mobile phone as it is the most personal and directly addressable device we carry, and its population already dwarfs that of all other directly addressable devices such as personal computers, digital TV and PDAs (personal digital assistants). Furthermore, we carry our mobile phone with us every day everywhere, something that we do with no other device except for the wristwatch. With direct contact costs similar to those of direct mail, the mobile phone offers a compelling marketing platform for advertising. Another unique benefit of the mobile advertisement is its direct call to action. A mobile ad can include a direct link to purchase or order, and such marketing was used for example in Hong Kong recently to sell out a concert by a popular pop band in a matter of 15 minutes. The registered fans received a mass-mailing of the offer to buy tickets and simply by clicking on the link they bought the tickets. The cost of such event marketing is a tiny fraction of what is usually used to promote rock bands on tour.

Mobile marketing can go much beyond that. Various services and content can be sponsored. For example, McDonald's could sponsor a game related to the current hit movie. Sponsorship with mobile services already exists, say, with free SMS text messaging services that are paid for by sponsorship, free daily news updates that are sponsored, free mobile games, etc.

A variation is the 'try me for free' business model that has been used to great success by the computer games industry. Give a small part of a game, or a 'first level' or trial version for free to anybody, and then those who get hooked on the game will be given the chance to purchase full games to play with all levels of the game or full functions of the game. This model is likely to be used increasingly with mobile games as the target playing user is more clearly identified by their previous game-playing history and trial offers can be made more accurately. The cost of delivering the games to precisely targeted potential users can be much less than the current costs of printing masses of trial gaming CD's that are handed out for free at various events and sales outlets.

With the various mobile advertising and sponsorship opportunities, the network operator and their partners will have a stake in the revenue-sharing arrangements and can generate income from such activities. Here the mobile operator must be careful not to step too far and start to annoy their customer base. Any advertising must be permission based, with the mobile-phone users expressly opting in to receive ads, not the other way around with users having to register if they don't want ads. The presumption must be that users don't want to be bothered. The ad campaign designers (and thus the owners of brands being promoted) must ensure that permission is granted and prove this to the mobile operators.

16.4 Revenue sharing

Revenue sharing is a new idea in telecoms, and as such it is still strongly evolving through trial and error. In most markets, mobile operators are not willing to share in traffic revenues, only in content revenues, although Norway has been innovative in also introducing some traffic revenue sharing tools. Most markets started off with content revenue sharing schemes of about 50–50, and only after I-Mode's 9–91 % split did the rest of the world start to migrate closer to the Japanese model. Today it seems that the typical offering that the mobile operator keeps somewhere around 20–30 % of the content revenues.

Revenue sharing is not that simple, however. There are several main factors which have an influence on determining the ratio of revenue sharing. The only common denominator with mobile service partnerships is that the mobile operator tends to be one party, but the other party or parties can be any one or several of content providers, application developers, system integrators, portals, banks, advertisers, etc. Therefore, this discussion will look at the factors from the angle of the mobile operator.

Exclusivity is the first determinant. If the service needs a mobile operator for the service to function, there is a considerable amount of exclusivity as most other operators, such as cable TV, fixed telecoms and ISPs, are not

Figure 16.3 Revenue sharing splits on a general level along the generic value chain

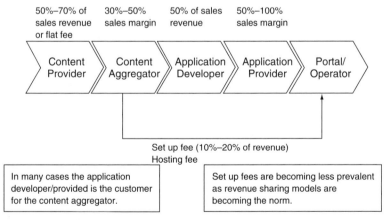

Source: BWCS Mobile Matrix, 2002

viable options. One example could be services delivered in real time to a moving car such as traffic congestion information. Even further, there can be exclusivity among mobile operators depending, for example, on a given technical standard, the availability of mobile phones and other terminals, geographic cellular coverage, etc. The fewer viable options there are, the more the mobile operator can expect to get as their share of revenue sharing.

The existing value chain or chains are a significant determinant. In many industries the delivery channel has introduced digital delivery and is likely to have the mobile Internet delivery option and its portion of the overall value chain relatively well defined. The mobile operator's share would be similar to what has already been introduced. If the industry itself is still in the early phases of conversion to digital delivery, then there is much more latitude for determining what the mobile operator's share is.

The extent of the value-add by both parties is of course also important. For the mobile operator, the main elements of direct value-adds are location information, access to user data, the micro-payment mechanism, and the ability to connect to direct communication systems, mostly voice and messaging. The more operator value-adds used in building the service, the bigger the operator's share can become. Equally the more value-add elements the partner brings into the service, such as software logic, security systems, exclusive copyrighted content, etc., the more that partner can expect to get as its share.

On-screen location is another key factor. The ideal position for accessing a mobile service is right at the welcoming screen on the mobile phone. With very little total available space on the phone screen, and with thousands of potential services vying for that location, the operator is in a strong position if the service is set to be close to the top of the screen. The service can also be built to be behind one or two keystrokes, which would have a similar effect. The higher the service is placed near the top of the opening screen, the more the operator can expect to get in its share of revenues.

Finally, the relative strengths of the brands are a factor. Brands will need to be considered always in the context of the given service and to its intended target audience. The stronger the content provider's brand, the weaker the network operator's revenue position is in relation to that given service proposition in its given target segment.

16.4.1 Revenue sharing levels

Three simple rules of thumb can be taken from the digitally converging world. In the credit card industry, credit-card companies tend to have commissions of between 2 and 4%. If the mobile operator offers no added benefit whatsoever from location information, subscriber data, or urgency/timeliness, and only provides the billing convenience and takes a risk replacing a credit card company, it is reasonable to assume that the operator would take at least the same amount as the credit-card company. As most credit-card companies have minimum payments, if the operator wants to offer micropayment options (payments with a value of less than 1 dollar), then the operator would be expected to keep more than the minimum that a credit-card company does.

The other extreme is splitting in half. While this may seem quite extreme, in many emerging digital services that need the operator's know-how and systems, for example developing a multi-user real-time network game, it is fair to assume that the operator gets a considerable part of the revenue due to its considerable involvement. For most combined service development, a 50:50 split would seem the other extreme. In this case the operator would provide location information, customer data and billing services to the partnership, as well as branding and location on its portal. Also the operator would take an active role in the system development or integration involved.

A third rule of thumb is the I-Mode example. DoCoMo's I-Mode service is widely quoted as taking a 9% revenue share from the value of the content or subscription to the official I-Mode sites. If no better initial benchmark exists, this can be a good starting point. A fair approach could be that starting from 9%, the two parties could examine what are the relative merits that either partner brings to the partnership, should the revenue share be more or less than with I-Mode. From that the relative extremes could be about 3% as the absolute minimum and 50% as the absolute maximum to be the part going to the operator. More realistically most service ideas could work on ranges from about 6% at the mimimum and about 35% at the maximum for the network operator. Pricing by the network operator would then be built to include elements like location information, bulk discounts, credit risk, etc., that would alter the basic price level but remain between the above ranges.

The content providers, application developers and any other parties wanting to get into revenue sharing need to look at their various options and consider what they would find acceptable with the operator. The content providers must be active in seeking the opinions from all players in their market, and use the bargaining power of playing one offer from one operator against the other to secure the best deal. Similarly, the network operators must remember that in their portfolio of thousands of services, there are rarely unique service offerings, and they too should be in contact with several candidates to find the preferred partners.

16.4.2 More money?

3G and advanced mobile services will almost always be built in partnerships and very often the business model will involve revenue-sharing in some form. Partnering is a completely new way of doing business for most mobile operators and they have to follow a steep learning curve to become good at it. Partnering will mean sharing profits, costs, customers, risks and information. We can take a lesson in how not to do it from this joke by comedian David Letterman: 'The CIA today announced they plan to cooperate more closely with the FBI; they just haven't told the FBI yet'.

In accordance with our principles of free enterprise and healthy competition, I'm going to ask you two to fight to the death for it.
— Monty Python's Flying Circus

17

Combatting Churn
Keeping Customers Loyal

When mobile operators were experiencing hypergrowth, they did not need to focus on customer loyalty, and 'churn' was seen as a trivial problem in contrast to getting all new customers connected. As the new-subscriber growth started to slow down, more realistic competitive rules came into play. At that point many mobile operators came to face churn as a new phenomenon. This chapter looks at churn and the related issue of customer

---

3G Marketing: Communities and Strategic Partnerships Tomi T. Ahonen, Timo Kasper and Sara Melkko
© 2004 John Wiley & Sons, Ltd   ISBN: 0-470-85100-7

loyalty. Just as segmentation could be considered as a basic principle underlying this whole book, churn management and related customer loyalty can be seen as the ultimate goal for long-term marketing success.

17.1 Basics of churn

The definition of a churned customer is one who has stopped using one service provider and has moved to its competitor. In some cases it is very easy to identify a churner, such as in cases of countries with number portability: if the customer, or one of your competitors, tells you to port the number of your subscription onto your competitor's service, that is a definite case of churn. A person who has left one operator to use another, has churned.

17.1.1 Who is a churner?

Most often however, it is not that obvious. Often in mobile telecoms a customer simply stops using one service provider. In such cases it is not clear if the customer has moved the traffic and business to your competitor. From the eyes of the mobile operator, this is not distinguishable from a customer that is just experiencing a temporary pause in using the service, for example, because the customer is short of money, the phone is broken, or the customer is travelling to a country where the phone does not work. The operator does not know whether someone does not feel like talking on the phone or has left the network altogether.

Prepaid accounts present a particular problem with identifying churning customers. A customer might leave a prepaid subscription to sit unused and start to use another service. As there often exists very limited information on the identity of prepaid customers, it would be difficult to verify whether an unused account is that of a rare user, or that the ending of use is due to natural causes such as illness or death, or the account was set up only for a short-term use such as that of a foreigner living briefly in the country, or it is actually the account of a churner who has moved onto a competitor.

The most difficult to identify are the hidden churners, who get another phone on a competing network and move most, but not all, of their traffic to the new network. Typically for such churners, they still receive some calls to their old number but rarely place calls on that subscription. They keep the old number because many know it, and for convenience reasons occasionally use that SIM (subscriber identity module) card, but for the majority of the calls they use their preferred provider. As second mobile phones are becoming ever more prevalent this will increasingly be the case. If you analyse the actual phone (and subscription) ownership in countries with penetrations over 90% such as Finland, Italy, Austria, Hong Kong and Israel, the penetration breaks down so that nearly 95% of the population aged 10–70 have one phone, and about half of the population aged 20–35 has two phones.

17.1.2 Why customers churn — three general reasons

The reasons people change service providers are varied. Some change for price/tariff reasons, others are unhappy or upset with their current provider. In countries where mobile phone handset subsidies are considerable, the prospect of gaining a newer phone handset is enticing customers to churn. Some customers change to get access to some services that the current provider does not offer while others churn to gain community benefits, such as having the family's phones on one network. There are three main reasons for churning, and these have dramatic impacts on the operator or service provider who is losing them and/or gaining them.

17.1.3 The joiner

The most common churner seen in the telecoms marketplace is the person we will call the 'joiner'. A joiner is attracted to the offering of a competitor. There is generally no need to have anything wrong with the current provider, only that something new by a competitor is attractive and that is a reason to switch providers. Joiners can be attracted to price offers, handset offers, new services or any particular reason why another providers offering is attractive.

The joiner is naturally interested in what other competitors offer. Therefore the joiner is one who is rarely satisfied by staying with any one provider for long. This is especially true of those joiners who are after strictly monetary benefits, i.e. joining a competitor because of better prices. The moment another competitor offers even better prices, the joiner will want to leave again. If some contract or other anchoring element is keeping the joiner, this will only postpone the inevitable, and many joiners will seek an excuse to leave anyway if attracted by a competitor's offering. Some estimates suggest that as many as 10% of the total population is of the type we could call a Joiner.

Table 17.1 Monthly churn rates in selected countries (%)

	1999	2000	2001	2002
Australia	1.9	1.8	2.1	2.2
Brazil	1.1	1.4	1.7	2.0
Canada	1.8	1.9	2.0	2.0
China	0.9	0.7	0.9	0.9
Finland	1.0	0.9	1.0	1.1
France	1.8	1.8	2.0	2.1
Germany	1.2	1.0	1.5	1.9
Hong Kong	4.0	6.0	5.5	4.0
Israel	0.4	0.5	0.6	0.8
Italy	1.0	1.0	1.3	1.3
Japan	2.5	2.2	1.9	1.6
Korea	2.3	2.3	2.2	2.3
Mexico	2.1	2.5	2.6	3.7
Netherlands	2.7	2.0	2.2	2.0
Russia	2.1	2.2	1.9	2.8
Singapore	1.7	1.9	2.0	1.8
South Africa	1.8	2.1	2.9	2.1
UK	2.8	1.8	1.9	2.1
USA	2.6	2.6	2.7	2.8

Morgan Stanley Wireless Matrix Dec 2002

17.1.4 The leaver

The second class of churner is the 'leaver'. The leaver does not leave because of something done by your competitor, the leaver is upset with *you* and wants to get away from you. *Any competitor* is better than you for the leaver at that point. The leaver probably would have stayed for a long time if not for something very specific that your company did or did not do that caused the urge to depart.

The good news with leavers is that the reason tends to be something that can be precisely identified. A bad customer-representative experience, a change in the service definition, or something like that, maybe even just one customer-service representative who is not really suited for that job. Odds are that if there is one customer who becomes a leaver because of that action, many others will follow. If you can identify the reason early enough and correct it, the leaver can be kept, and even if not, at least identifying the reason allows retention of other potential leavers.

17.1.5 The changer

The last type of churner is the 'changer'. A changer leaves not because of anything you did, or what the competition offered, but for their own internal reasons. It can be moving residence, for example, or a change in marital status, or for a business customer a change of a manager who wants to try something new, etc. It can even be a life philosophy along the lines that change is good, with some people perpetually wanting to experiment with something new and thus never remaining loyal forever.

Changers will probably leave almost no matter what you do or don't do. Changers will also leave your competitor, so you will naturally be gaining your share of changers over time.

17.1.6 Selecting customers to target

When determining which existing customers to target for churn reduction or customer loyalty programmes, the first criterion must be profitability. For any customers who are unprofitable to you, the only real action can be either to try to turn them into profitable customers or to get rid of them. You should keep in mind that no churn reduction or loyalty programmes help in turning a currently unprofitable customer into a profitable one; on the contrary, such programmes add further costs to a loss-making customer. If you cannot turn a non-profitable customer into a profitable customer such as by raising tariffs, adding new fees, or getting the customer to use services that bring profits, then the best thing to do with such customers is to give them to your competitors.

As to customers who are joiners, these tend to want to change often anyway. The cost-effectiveness of any loyalty programme must be measured not against the profitability of the average customer over its average lifetime, but rather against the profitability of a joiner-type customer and their much shorter time as your customer. If the loyalty programme or churn reduction plan is cost effective in extending the short-term stay of a joiner, it is worth doing, but if the programme does not pay itself back by the longer stay of the joiner, then it is not worth doing, and the joiner should be allowed to go. Don't build loyalty programmes on the premise of locking in joiners, remember joiners are only yours for a short duration anyway.

Regarding the leavers, there is clear reason totally within your company control why these customers left. It is most likely more cost effective to deal with the true cause of their leaving rather than trying to entice them to stay with a loyalty programme, while at the same time annoying them with continued action (or lack of action) that they detest. Critical to spotting your internal actions that caused leaving, is to monitor what your customers truly think, actively to encourage complaints, to reward complainers, and most importantly to deal with any causes that provoke leavers.

The changers are likely to change anyway, irrespective of what you do, and a loyalty programme will usually not be able to prevent a changer from churning.

17.1.7 Stayers

Those customers who are not likely to churn are stayers. In most market conditions the total number of stayers is much greater than all other types of churner combined. For customers who are not current candidates to churn, a loyalty programme can build stronger ties and reinforce the willingness to stay.

It should be noted that incumbent operators naturally have a proportionately larger portion of stayers than the new entrants. This is because stayers spend a very long time with one operator or service provider and rarely bother to churn. Newcomers have the highest proportion of joiners because joiners change operators often and are like nomads, moving in herds from one operator to the next, often returning only to leave again.

The leaver is the ideal type of churner to acquire. The leaver has no inherent desire to change operators, and would have left the last operator because that operator had become unbearable for some reason. The leaver is very likely to try to find any justification in their next operator to remain with them. If you treat a leaver's reason for leaving competitor, and give them good service, a leaver is likely to remain with you a very long time as a satisfied customer.

17.2 Churn is good — targeting competitors' customers

The most powerful way to attract customers from a competitor is to identify both weaknesses in the competitor and customers for whom that weakness is a significant factor. If such a customer can simultaneously be turned into a leaver and be convinced that your service is worth coming to, you gain

Figure 17.1 Mobile phone ownership in the UK by age group (iSociety Mobile UK March 2003)

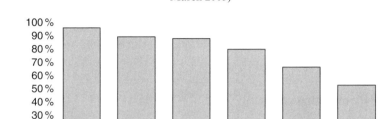

a very worthy customer. The key to identifying prospective churners and the conditions under which given segments might churn is naturally market intelligence. You will need continuously to track what the customers of your competitors think of their current providers and try to isolate what might make them move. The only way to isolate the opportunities is to track customer satisfaction in a regular and formalized way, so that trend data can be developed. Only then can you see when opportunities emerge for given segments. On a practical level, if you work for a network operator in a marketing management type of position, make sure you have a subscription to each of your competitors' services and use them regularly to monitor how they compare to your own. It is the most cost-effective early radar for you to keep an eye on your competition.

17.3 Churn is bad — don't let valuable customers churn

We have shown that there are extremely costly customers. Of course there are also extremely profitable customers. A widely reported study by Bell Canada found that 14% of customers bring in 53% of the revenues. In general in telecoms, the conventional wisdom suggests that the '80–20' rule seems to apply in that a small proportion, 20% of customers, brings in the majority of the revenues, 80%. The Bell Canada study supports this kind of division even though the exact numbers do not coincide with 80–20.

It is important to remember that a high-revenue customer is not necessarily a high-profit customer. When ranking customers for their value, several elements should be considered, including projected lifetime value of the customer relationship, its projected revenues and profits, and any reference value the customer may bring. With residential and small business customers, the network effects of customers who bring in other customers can be an overriding factor in determining which is the most valuable customer.

17.4 Combatting churn

There are numerous tools for addressing churn specifically, with some working better than others. In the long run, the only way to hold onto a customer is to keep the customer satisfied. Over the shorter term there are several tools that can be considered and merit some discussion in this chapter.

17.4.1 Price as the weapon

The most common marketing tool in telecoms has traditionally been 'lets offer it with a lower price'. At first sight lowering the price might seem like an inexpensive way to acquire a competitor's customer. However, with this strategy the attracted target groups are the typical joiners and other bargain hunters. This means, as outlined earlier, that you get highly volatile and very disloyal customers, who keep on searching for the cheapest service provider and lowest price.

Still, many inexperienced service providers will try to do their marketing by price. A low price seems like the simplest argument for a change and intuitively it sounds like the most important decision criterion for the customer. The reader of this book must keep in mind that price is only one of the various marketing tools and a targeted approach with the right *mix* of marketing tools can yield better success with more loyal customers and with better cost efficiency. The important factor is, of course, that all of this requires real work and a solid understanding of marketing. It is *easy* to adjust the price. It takes *know-how* and *hard work* to identify what is truly needed to attract (or keep) a customer.

When low-price players emerge in the market these are most often likely to be on alternate technologies or as MVNOs (mobile virtual network operators) or service providers typically intending to compete only in well defined niche markets. The low-price strategy may also be selected by the smallest of the major network players. If your competitor starts to slash prices it is not advisable to compete directly. Small service providers may find innovative business models focused on certain niches and often can live with more meagre profit margins than can major network operators with a lot of infrastructure and R&D waiting to be financed. The problem in using price cuts to answer niche market players is that, invariably, the bigger operator's price cuts affect larger segments of the total customer base, dropping prices for many who would not be attracted to the niche player's arguments.

The main question therefore becomes: Why should one eat one's own profitability for disloyal customers who are not very attractive in the end? Entering a price war is always a bad option. Of course we are not advocating unreasonably high price levels either. If an operator's service prices are clearly too high, sooner or later they will have to lower their price levels to a standard that is more in line with the overall market level. Even stayers will not accept everything. The trick is to find the optimum point, which should be somewhere lower than the threshold of pain where even a loyal customer would prefer the trouble of changing the operator for the price's sake. We discussed pricing in Chapter 14.

17.4.2 Technical barriers and churn

An engineering approach to churn is to consider technical means to 'lock' the customer. Some of the most common such means are, for example, number portability, either the total lack of it, or by making it cumbersome and expensive. Another is non-standard components or systems, such as non-standard SIM (subscriber identitity module) cards, or non-standard USIMs (UMTS subscriber identity module) in 3G. Also the walled garden idea of a mobile portal is another technical barrier.

The appeal of a technical lock to customers is alluring, especially to such engineers who are not interested in discovering how to *attract* customers. The term 'lock' is appropriate to describe what is intended: the mobile operator wants to imprison the customers. The reader of this book can understand that in our view, any moderately free market will punish such behaviour in the long run. Some competitors will provide solutions that circumvent or avoid the technical means of binding customers and after that those operators who were seen as abusing the technical means will be punished by the marketplace. Customers will churn if they want. The only way to prevent churn in the long term is to make customers *want* to stay — and that is what this book is all about.

17.5 Number portability

One of the strongest anchors for keeping customers and preventing churn is the mobile phone number. It does not matter much in countries where number portability has been implemented with minimal trouble and cost to the phone user, but in many countries just the delay and aggravation caused by going through the number portability process is enough to deter anyone leaving a network and joining another.

If number portability is not a viable option, then for many users a mobile phone number commits that person to the network. The trouble of changing numbers in itself is not such a strong deterrent, but the fact that everybody needs to be informed of the changed number makes the activity undesirable. For business customers, and all who depend on the mobile phone for work, etc., the change of numbers could mean lost business. With business customers the printing of marketing literature, from business cards and letterheads to brochures, etc., can be another cost that suggests changing numbers is not desirable.

17.5.1 Changing numbers

The mobile subscription is nowadays a person's most crucial link to his/her social world. When thinking about changing a mobile operator to another two main concerns arise: (i) how can I inform all my contacts in case my number changes, and (ii) how to copy all my contacts from one SIM card to another without typing everything in once again? Mobile operators need to consider these issues both in holding onto customers and when trying to convince customers from other networks to change.

One solution to diminsh the stress and cost of changing numbers and networks is a combined SIM card conversion and free change notification package. It works so that the mobile operator offers to copy the old SIM card's content to the new one, and also to grant one free SMS-message to every number stored on the old SIM card. This will allow the customer to inform all those friends who were important enough to be stored on the old SIM card of the change in the number. From a financial point of view it is important to bear in mind the costs of these services as they add to the initial acquisition costs of new subscribers along with the other costs such as marketing, handset subsidies, introductory prices, free bundles of minutes and/or sms/mms, etc.

17.6 Loyalty programmes

A tool created for the purpose of keeping customers with one service provider rather than switching to a competitor, is the loyalty programme. Loyalty programmes tend to be modeled along the ideas pioneered by the airline industry, collecting airline miles and exchanging them for awards. Many telecoms operators have introduced loyalty programmes and one of the early pioneers was Omnitel Pronto in Italy. The initial programme was aimed only at post-pay customers but since then the programme has been expanded to include prepay customers as well. Typical of loyalty programmes is that the points collected can be exchanged for telecoms products, such as free messages or free upgrades of the handset, etc., but often include awards with partner companies in non-telecoms services and products as well.

Points can be collected for regular service use, such as minutes called and SMS messages sent. Bonus or campaign points can be given when new services are introduced to promote the early adoption of new services. Anniversary points can be awarded at the anniversary of the contract and in this way to reward long term loyalty. The variations and complexity of the points programme are only limited by the imagination, but practical limits are brought about by the consumers, they will not want the programme to be too difficult to understand.

Usually operators will try to get partners from outside the telecoms industry to join in and give points for purchases. The single most popular partner tends to be the dominant airline of that country, such as Alitalia in Italy or Air Canada in Canada, etc. Callers are particularly fond of the concept of gaining airline miles from using the mobile phone. Airlines, however, are becoming ever more aware of the value of their miles and are starting to charge higher rates for giving airline miles for partners and some telecoms operators have stopped giving miles. The mobile operator should always include lots of its own services as awards, at very 'competitive' prices when charged by points. It is always less costly to offer own services rather than having to pay for those provided by other partners.

17.7 Handset subsidies

As we already have said briefly in the chapter on handsets (Chapter 7, Terminals), the amazing thing about handset subsidies is that it varies so dramatically, from one extreme to the other. Most mobile telecoms marketing professionals have experienced only one type of market and the difference can be mind boggling.

The first issue that all mobile telecoms marketing people should know, is that subsidies can be radically different in any given market. In most markets mobile phone handsets are subsidized. In extreme cases phones are given for free or for a token fee such as 1 dollar. For readers who work in countries with no subsidies, the market rules and dynamics of dramatically subsidized markets can be quite overwhelming. In markets with subsidies the handset subsidy is the single biggest cost element of a customer lifetime. It is more costly than all other marketing activites combined. It is more costly for most average users in the average lifetime of the customer, than the actual network resources used. Because of subsidies, it becomes a delicate issue in ensuring that the overall customer proposition is profitable.

For those who work in markets with subsidies, it may come as a shock that of the world's leading markets that have been at the very forefront of mobile phone penetration for the past ten years, in Italy and Finland — both with subscription penetrations well beyond 90% and still rising — there are no subsidies whatsoever. All consumers pay full retail prices for their mobile phones. This has in no way deterred them from buying ever more expensive and feature-laden handsets and replacing them at roughly the same rate as the rest of Europe. Recently Korea joined the subsidy-free market, and we expect many more countries to follow during this decade. For those who are lucky enough to work in markets without subsidies, you can skip this part of the book.

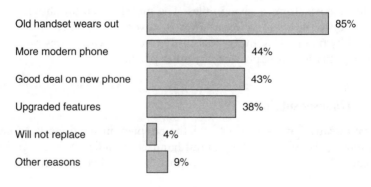

Figure 17.2 What factors cause UK users to replace mobile phones (Enders Analysis October 2003). Multiple replies were allowed in the survey

17.7.1 What makes subsidies so damaging?

Whenever a market has subsidies, the subsidy is a big motivating factor in consumer behaviour, often being the biggest single factor. Subsidies often trigger market distortions, which can be quite massive in provoking actions that would not take place otherwise.

Handset subsidies were initially offered when the early cellular phones were perceived as being very expensive telephone gadgets, especially when compared with fixed line phones. The handset-subsidy idea follows many similar business models where an enabling component is priced below profit with the aim of making that up with another part. Famous examples include the traditional men's razor blade business (before the advent of the disposable razor) where selling the razor was not profitable but selling the razor blades was. The idea has also found its way into the current video-game console, again where the PlayStation 2 or Xbox may be non-profitable but individual games are very profitable as a business.

Mobile phone subsidies have now distorted the behaviour of regular customers. The mainstream theories of economics and marketing assume that masses of customers behave like a rational person. Subsidies introduce a significant element of irrational behaviour, drastically complicating and frustrating the efforts of the marketing executives at the mobile operators. Subsidies were, of course, intended to hook customers to a given operator, but after subsidies become the norm for the market, and with ever newer handset models with more advanced features, they foster a culture of switching. To show by analogy, the larger the gifts (value of the subsidies) the more gift hunters are attracted, and this results in increasingly more disloyal customers. Customers have been *conditioned* to *expect* free or extremely low-cost phones. The mobile operator who now has the customer will have hard time in convincing that customers not to change to a rival who offers a subsidy, thus forcing the mobile operator to keep buying back its own customers by giving them ever more expensive handsets.

With handsets being tied to subscriptions, the following negative effects emerge: (i) handset manufacturers cannot convey quality or worth aspects by price, and (ii) handsets will move to the centre of the discussion. By turning attention away from the actual services, such as the subscription and advanced wireless services, the operators do themselves a massive disservice.

As we said before, handset subsidies are inherently bad in any mobile telecoms market at all times. They are very difficult to get rid of once they have been introduced into a market, but it can be done. Subsidies are wholly unnecessary for a vibrant mobile telecoms market, as proved by Italy and Finland. As we have seen, subsidies trigger behaviour where the users do not follow reason. In sum, the subsidy situation presents a classic 'prisoner's dilemma' where often the industry understands that the subsidies are bad, but any single operator cannot find it in its interest to stop the practice unilaterally. Here the operators can lobby the regulator or try to find industry consensus to steer away from subsidies. When doing this, operators must be mindful of not being accused of collusion, so educating the market to work closely with the regulator is critical. It will not be an easy road, but in such an uncertain world as 3G, perhaps this is one uncertainty factor that the industry should resolve to its benefit. We strongly urge the operators in any market to find a lawful but swift road to bringing a reduction and eventual end to subsidies. The Korean example is proof that it can be done.

17.8 From techniques of authentification to identity

Show me your mobile phone and I'll show you who you are. Mobile devices have experienced a mutation from technical gadget to a highly personal tool. There is a whole industry around manifesting your uniqueness through your handset, starting with the many different models, interchangeable covers, add-on keyboards, ringing tones, on-screen logos, etc.

The mobile device has become a symbol for our very existence and our daily lives. Gone are the days when the worst thing to be forgotten when going to work was the wallet. Never mind the wallet — but not without the handset! The total dependency on the little mobile friend — called 'Teddy' as in teddy bear in Sweden — is blatant when one goes though it for the first time: the annihilating feeling of being totally cut out of the world without the possibility of communicating. The feeling of being deprived of social contacts with our communities and the nagging feeling of missing something particularly important both through the calendaring function and urgent messages arriving via SMS text messaging.

Mobile payment makes the handset actually superior to the wallet. Even if you have no money, in countries where m-commerce is commonplace you can survive by paying with the mobile phone. For example, in Helsinki it is possible to eat candy bars, drink sodas from vending machines, and pay for transportation all by mobile phone, giving much of the very basics one needs to survive and get back home. Besides isn't the invitation 'can I sms you a drink?' much more charming than it's equivalent using the more commonplace term 'buy'?

Table 17.2 Planned and unplanned tasks

30-Minute tasks	30-Second tasks
Planned	Unplanned
Sitting	Standing/walking
Big display	Tiny display
Keyboard and mouse	Keypad and stylus
Create info	Consume info
E-mail	SMS
Personal Computer	Mobile Phone

The more the device — or more precisely the subscription — becomes the means of authentification with m-payment and other applications, the more users will identify with this means of technology. Mobile operators have to make sure that the identification will not be the sole benefit of the handsets. Certainly people tend to notice tangible products more readily than abstract services. Nevertheless, if an operator succeeds in shifting even a small part of the customers' identification potential from the device to the operator's services or even the operator itself the gain will be considerable in the area of loyalty.

17.9 Back to the brand

When talking about communications we are back to branding. Why some people swear by the superiority of one carbonated sugar beverage over another? Of course, you cannot entirely compare a mobile subscription to Pepsi Cola. Compared with the purchase of convenience goods, a mobile subscription is more muted in its market behaviour. Changes in market loyalties are more subdued due to rational choice, coverage and service levels, the linkage of subscriptions to handsets, all meaning longer commitment. In many countries, mobile subsciptions have minimum durations, most commonly 12, 24 or even 36 months. Therefore, increasingly, it matters ever more which company you have most faith in, think suits you best and communicates in the language you understand. Again, a company's real value is mirrored best in the people's minds and hearts. An appealing message or image is a good enticement; if not necessarily the only reason, it most certainly enforces changes.

Uncovering killer applications is of course the best short-term magnet for attracting new business. As we know, killer applications are not invented every day and even when they are, invariably they will be copied by the competition, thus limiting the benefits to acquiring (or keeping) customers over the longer term. None the less, a company has to recognize whether or not a service in their actual portfolio has the potential to be a real killer application for a certain market segment already, and if so, to start communicating this. Often, a killer product is, for a large part, right communication to the right people.

When talking about your competitor's customers you have to know your customer well. As we discussed in the early chapters to this book, this is achieved by thorough business intelligence and segmentation. The strategies could then be as follows: either to differentiate from competitors and attempt to create a unique service proposition; or try to benchmark the competition, but outperform it in some way relevant to the customer. Investing in market research provides data on what the competitor's customers appreciate. The need is to have a systematic way of collecting standardized data that can be tracked over time. This gives the chance to perform trend analysis as trend analysis gives input for service development efforts.

17.10 Community think

Mankind has always moved in groups or like animals, herds, and this is still true today. What defines a clan or its bigger form, a people, is their common culture, i.e. the common way of communicating and behaving with one another. Broken down into our day to day information technology, the communication behaviour means — in an abstract form — that every person belongs to a certain group of people having the same language, interests, humour, and means and devices of staying in touch. In the mobile telecoms context it can even mean that a community has the same mobile service provider.

The mobile telecoms industry has studied the concept of communities for the last few years and a lot of new group behaviour has been observed that manifests itself via the mobile phone. We will group our contacts, colleagues, family and friends differently as we classify them onto our phones. We may use nicknames for some. We may assign different ringing tones to our nearest and dearest, such as our immediate family. We choose to communicate with different communities in different ways, some we call, others we send SMS text messages, yet others we actively ignore.

Combatting Churn

Mobile operators, who are starting to understand these mass dynamics and more or less even the peer pressures involved, are looking at communities as a very powerful means of binding people together and — as a group — to themselves as their favoured service provider. Individuals might change their minds and directions quickly and thus also may suddenly change service providers. Not so with a greater amount of people — the trotting of a herd is a steady flow. Thus, groups of people — communities of family members, company employees, friends — consitute a very important and loyal customer segment or segments. Operators are better off feeding and attracting these groups by granting special conditions for them. Examples for community services start from special prices for your most popular, i.e. your community's registered numbers, then special services that tailor to them, such as group calendars, chat boards, etc.

When choosing an operator often the first thing customers do is to ask people they know which service or operator they are using. This works as a referral and it is usually less expensive to place calls on the same network. Individuals should be rewarded for attracting new subscribers onto the same network. Offers such as 'member get member' with various bonuses and group offers may enhance the changing of whole groups of people from one service provider to another or to get someone who is outside the community to return to be with the others. A steady diet of 'it is so expensive to call your network' is gruelling in the long run. The reader should also bear in mind that, increasingly, users will start to have multiple subscriptions — at least in most of the world except perhaps North America with its archaic receiving-party-pays billing concept — and this second subscription selection can very strongly be influenced by the community aspect.

The operator should also recognize that the behaviour on a second subscription will be quite different from that on the primary subscription — but no inherent reason exists why the two could not flip over. Clever marketing activities can convince a new user to place the majority of the telecoms and mobile data traffic with the new second subscription, effectively making that the primary subscription. Also readers of this book will recognize again that what is good for one player is bad for the other. Churn always is a two-edged sword. If one of your competitors is starting to attract your customers to second subscriptions, then beware, these may gradually shift their traffic to the new service. Worst of all, your systems might not detect it, if there is still also traffic on your network. You would be losing the battle without ever knowing about it.

Communities are usually strongly influenced by key leader individuals. These should be identified and targeted to become mobile service evangelists. And *voilà*: there we have a major argument to be first on the 3G markets. Innovators are often opinion leaders and as such, play an important role in convincing their communithy to join them. It is harder to alienate a customer group once they have made their choice. Logically, a big ship changes direction more slowly than a small boat.

Community thinking is an accelerator for a product or a service and has its own momentum. One of the most popular examples is Harley Davidson — you do not buy a mere product — you buy an entire way of life and directly obtain all the aspects associated with the brand. When you come to think of it — there is no top of the mind brand in mobile communications let alone 3G globally yet. The potential is there to create a strong brand around the very characteristic that make us human — communications.

17.10.1 Customer intelligence and churn

Simply put, the better you know your customer, the more you know about his preferences, and the more able you are to serve the needs, the greater the loyalty. As we saw in the customer intelligence discussion earlier in this book, modern customer intelligence needs powerful customer relationship management (CRM) tools and methods. Ideally, the customer's whole history and relevant data is collected, sorted, analysed, clustered, portrayed, and prioritized in order to make good management decisions about customers and the business. While dealing with similar content, the CRM tool is quite different to the billing (and charging) system. Billing systems should generate data to the CRM but the CRM should also receive input from other sources. A CRM tool should be able to answer the following types of question:

- How was the customer acquired (what was the catch)?
- What services does the customer use?
- How is the customer classified in the segmentation?
- Is the customer profitable?
- Is there a typical pattern of customer behaviour?
- What is the cross-selling potential?
- What activation campaigns have been used?
- Are there similarities between other segments?
- Have there been complaints, special treatments?

Of course, CRM tools might vary substantially in their extent and in the information gathered when it comes to consumer (mass) markets and business customers. The business-user CRM would also identify decision makers and key influencing contacts and include a full history of sales visits, service technician histories, information on in-house systems in use such as PABXs (private automated branch exchanges) and IT systems, etc.

All through this book the reader should bear in mind that holding on to existing customers is less expensive than trying to attract new customers. When used properly, a good customer intelligence process and system will save money by identifying customer needs.

17.10.2 Keeping customers happy

After identifying what it is that a customer wants, or may want, according to the segmentation insight, the next step is communicating with the customer(s). Here it is better to go through the trouble of asking the customer for his preferred means of being informed. Some will prefer to receive information with the written phone bill, others may want it as e-mail notifications, still others may want SMS text messaging notification, etc.

Here the operator must tread very carefully. There is a fine line between useful information and spamming. The customer's willingness to receive information via the mobile phone should be treated very cautiously. What may initially seem like useful information may soon be considered spam. The operator must be very diligent in limiting the use of more intrusive media. What is perfectly appropriate advertising of new upcoming services as a billing leaflet insert, will become annoying in e-mail and infuriating in SMS text messaging. Very strong internal guidelines should be set on the absolute maximum levels of notices that may be sent on any of the media, and the internal new service managers and segment marketing managers, etc., should then present a strong case for why their service should be mentioned in the one marketing campaign rather than another.

The same applies to selling customer data or giving partners access to it. Among an operator's most valuable asset is the deep insight into its base of customers and the direct contact to each customer at practically all times. This will grow increasingly important during this decade, as intrusions into privacy will become ever more prolific ranging from cookies on Internet websites to increased security checks relating to terrorist threats to added surveillance to the global integration of databases. The mobile operator must treat customer information as a precious and near-unique natural resource to which it has an exclusive right to mine and use.

It could mean direct self-cannibalization if customer contact and data is exploited too harshly or too often. The mobile operator must bear in mind that its customers carry the mobile phone with them at all times and treat messages and contacts coming through it as the most personal of all communication. Getting spammed by all kinds of unknown entities will create extreme bad will that can result in anti-spam actions through legislation and regulation, and definitely provoke change of operators. Again, with the advent of multiple subscriptions, it can be very difficult to detect when a customer has actually churned, as the existing subscription may still occasionally receive calls, messages, and even be used at times.

17.11   An end to churn

Churn was not an issue when the industry was growing at double digit rates. As the mobile telecoms marketplace becomes more conventional as a competitive environment, so too will regular rules of competition come into play. Understanding what makes a customer loyal, what causes customers to leave, and which customers should be enticed to stay, are keys to a profitable customer base. Understanding and managing churn will become a key to success in mobile telecoms.

Customer satisfaction and managing churn will be perhaps the biggest competitive challenge in 3G. Success with it will determine to a very large degree the eventual winners of the 3G market. Successful churn management will require professional marketing consistently performed, with lessons learned and applied over time. The reader should keep in mind, however, that marketing is as much an art as it is a science. As such there has to be room for creativity and innovation. Make sure that you nurture the creative elements. As Pablo Picasso said: 'Every child is an artist. The problem is how to remain an artist once he grows up'.

Hyvin suunniteltu on puoliksi tehty. (Something well planned is half completed)
— Finnish proverb

18

Marketing Plan
How to do it

This book has discussed the major aspects of marketing as they relate to mobile telecoms, with a focus on the marketing activities of the mobile operator. In the professional execution of marketing within a modern company, the most important single tool is the marketing plan. Marketing activities should be well considered, with clearly indicated purposes, tools, roles, resources and timetables, in a documented plan. This chapter will briefly discuss the marketing plan.

18.1 Business, marketing, advertising plans

As with many of the chapters of this book, the idea of the 'marketing plan' can have many meanings to many people. We want to identify the marketing plan within the context of other related planning tools to help clarify exactly what is and what is not a marketing plan. Consistent with the rest of the book, we use the definition of marketing as being all activities used to bring products and services to the market, not only advertising and promotion.

18.1.1 Business plans

In a modern professionally run company, a business plan precedes the marketing plan. The business plan follows from strategic plans and goals. A business plan is a document that identifies how a given product, service, or area of products or services, of a company will achieve business aims. Usually the business aim is profit although it may in some cases be a market-visibility goal, a market-presence goal, or even a performance goal in relation to other players in a market, etc. The key defining elements of any business plan are definition of intended markets that are to be addressed, the pricing levels used, amount of business generated, the resulting revenues, the related fixed and variable costs incurred, and the profit/loss incurred. A business plan will include a budget for the whole business activity.

A business plan sets marketing objectives, such as intended market shares, usage levels, marketing budget cost projections, etc. The purpose of the business plan is to identify and analyse a significant business opportunity and set specific goals that need to be achieved for capitalizing on that business opportunity. However, a business plan does not go into any detail about how any given marketing element is determined or executed. A business plan will defer to a marketing plan.

18.1.2 Marketing plans

A marketing plan takes intended marketing performance goals such as market share as givens from the business plan, and then proceeds to identify the individual marketing activities that need to be accomplished so that the goals can be met. The marketing plan will determine which marketing activities are needed, identify timing, costs, resources, and responsibilities of each. The marketing plan will not go into detail about the production costs of the products or services being delivered, unless given marketing activities cause incremental costs. The marketing plan is not intended to provide business justification and go into detail about profits and losses of any business proposition.

Marketing plans cover all relevant marketing activities that are needed to fulfill the objectives of the marketing side of the business plan. As such, the marketing plan can cover any issues that have been discussed in this book, from customer segmentation to product development to brands, promotion, sales, distribution, etc. The marketing plan will also set specific goals for promotion or advertising plans and, where relevant, goals for sales plans, product development plans, etc. The advertising, PR and related promotion activities are often planned together with advertising agencies or PR agencies. They may want to develop what they call a marketing plan for their activities. It is important to develop a plan, but for the purposes of this book, it is very important to note that this kind of planning is at best a promotion plan, or can be an advertising plan or PR plan. While it will be important that such a plan covers marketing objectives, it should not supersede the existing marketing plan. The promotion plan should be completed only after the marketing plan has been approved.

18.1.3 Hierarchical nature of plans

The order of the plans in chronological order and in managerial attention is: strategic plan; business plan; marketing plan; promotion (or advertising) plan. It is very important that this order is maintained, and that the plans follow on from each other. Without a strategic direction, individual business plans can be unsupportive of each other, split company resources and efforts and, in the worst case, be at cross purposes. Strategic plans usually run for several years and have annual updates.

Only after the strategy is set can business plans be developed. It is common to have several business plans for the strategic product lines and services. In some companies every significant product or service must have an approved business plan. Any business plan must be completely in line with the strategy. Business plans can run from a year to a few years, but not longer than the strategic plan.

After the business plan is set, a marketing plan can be developed. There is usually only one marketing plan for any business opportunity. Marketing plans are always to a set timetable, with clearly defined goals that are to be achieved with the plan, such as a market share, a level of market awareness,

Figure 18.1 Time spans of marketing-related plans

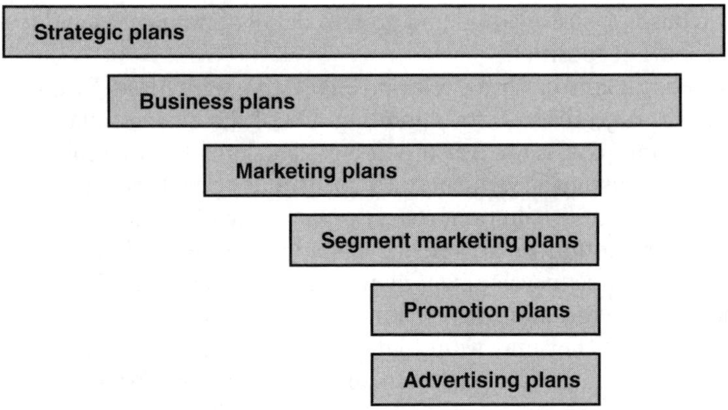

- Strategic plans
- Business plans
- Marketing plans
- Segment marketing plans
- Promotion plans
- Advertising plans

etc. Marketing plans usually run no longer than a year and may be as short as 3 months. If a business plan isolates multiple opportunities for example for a series of products or services, it is possible to develop multiple marketing plans to support one business plan. If so, these separate marketing initiatives must be identified in, and fully in line with, the business plan. Marketing plans must have a specified body or executive outside of those executing the plan, who will review the accomplishments of the responsible managers at the end of the marketing plan period. The review will compare the goals stated in the original approved plan to the achieved levels after that period.

18.1.4 Segment marketing plans

Within a marketing plan it is possible to develop focused marketing efforts in a series of 'segment marketing plans', usually run by segment marketing managers. These would be subsets of the overall marketing plan, and will identify targeted activities aimed at given segments. For example, if an overall marketing plan intends to capture 23% of the 3G market over the next year, three segment marketing plans could be developed for the most critical identified target segments, such as innovator business customers, early adopter residential customers and frequently travelling business executives. The three segment marketing plans and their individual goals would be specified in the overall marketing plan, while individual allocations of marketing resources would be identified in the segment marketing plans. Segment marketing plans are typically for half a year or less.

Only after a marketing plan, or related segment marketing plan, is completed and approved, can a promotion, advertising or PR plan be developed. If this activity is done with an outside agency, it can be very useful to develop this plan with the outside experts. They should be given access to the current approved marketing plan, and the goals for the promotion, advertising or PR plan should be taked directly from the marketing plan or segment marketing plan.

18.2 Marketing plan outline

There is no perfect outline for a marketing plan that will cover all eventualities. In mobile telecoms, however, a general set of commonly relevant marketing activities can be provided and thus an outline for a 'major' marketing campaign can be identified. We have listed these as the outline for the draft marketing plan and suggest that this be used as the starting point, and any non-relevant items be removed from the actual marketing plan that is to be developed for any given use.

The marketing plan outline is as follows:

(1) executive summary (if plan is beyond 10 pages in length);
(2) objectives of the plan, including references to existing business plan;
(3) duration of plan;
(4) deliverable targets with exact metrics, dates and responsibilities;
(5) target markets (if relevant, e.g. pan-European operator might list target countries that the plan is for);
(6) target segment(s), including references to subsequent segment marketing plans if any;
(7) competitive environment, including references to any competitor analysis that may be relevant;
(8) branding;
(9) service offering and bundle, including references to any R&D, product development and/or product management plans that may be relevant;
(10) terminal impacts; including references to any vendor specifications, delivery commitments and cooperation plans as relevant;
(11) distribution channels, including any related channel management plans as relevant;
(12) pricing, including any promo discount pricing;

(13) promotion including publicity, PR, advertising, sponsorship, product placement, to include reference to any subsequent promotion, advertising PR or other such plans, and to relevant corporate promo guidelines;
(14) sales and sales support including any sales campaigns, sales tool kits, argumentation guides, calling centre scripts, etc., to include reference to any subsequent sales plans, contests, etc.;
(15) portal use if relevant;
(16) community aspects if relevant including member-get-member;
(17) provisioning of service;
(18) customer care and billing;
(19) customer satisfaction follow-up;
(20) marketing materials including any brochures, price lists, white papers, user guides etc.; also any marketing gimmicks and give-aways;
(21) sales support materials;
(22) internal marketing;
(23) training;
(24) staff to execute the plan, including manager in charge and any organizational structure needed for the plan;
(25) resources to be consumed;
(26) responsibilities of plan approval, oversight and review;
(27) evaluation of plan execution (by outside entity against originally approved goals);
(28) budget;
(29) schedule of major activities.

Any such plan is a draft until it is approved. Once it is approved the approval date and approving entity should also be indicated in the plan. The above list is not complete but should serve as a check-list and assist in selecting the relevant items. As the bare minimum a marketing plan should have the following:

- objectives of the plan;
- duration of plan;
- deliverable targets with exact metrics, dates and responsibilities;
- target segment(s);
- service offering and bundle;
- pricing;
- promotion;
- sales and sales support;
- provisioning of service;

- customer care and billing;
- customer satisfaction follow-up;
- marketing materials;
- sales support materials;
- internal marketing;
- staff to execute the plan;
- resources to be consumed;
- responsibilities of plan approval, oversight and review;
- evaluation of plan execution;
- budget;
- schedule of major activities.

The information in the marketing plan is necessary for all those who are actively involved in implementing the plan. It is also important for those departments, units, and individuals who are affected by the plan. Marketing plans or parts thereof may also be shared with close partners in campaigns where those partners have an interest.

It is very important that marketing activites are conducted by professional business methods. This means that plans are prepared, approved, the progress is monitored, and the results collected and compared with the plan. Only through this method can lessons be learned and the marketing activities and processes be improved over time.

18.2.1 Plan ahead

This was a short overview of the planning process and how a marketing plan fits with it. We have listed numerous optional items for those marketing plans that are broad and comprehensive. For those who wonder which to select from our list, we can only say, use your best judgement and try to cover everything that is relevant. As J.R.R Tolkien wrote: 'It does not do to leave the dragon out of your calculations, if you live near him'.

The greatest and noblest pleasure which men can have in this world is to discover new truths; the next is to shake off old prejudices.
— Frederick the Great of Prussia

19

Postscript
Entering a New Era in Telecoms Marketing

Telecommunications as an industry has existed for much over 100 years in a stable and predictable state and was, until a few years ago, under protection of monopoly. Liberalization of telecoms did bring about new entrants into fixed telecoms, Internet and datacoms businesses and also into cellular telecoms. The first new entrants were mostly people who had

---

3G Marketing: Communities and Strategic Partnerships   Tomi T. Ahonen, Timo Kasper and Sara Melkko
© 2004 John Wiley & Sons, Ltd   ISBN: 0-470-85100-7

strong telecoms backgrounds and thus the steps to competition were not overly aggressive.

The Internet business benefitted from the rapid growth rate and fierce competitiveness of the PC business. Many countries went through the predictable cycles of emerging from monopoly, with literally hundreds of start-ups, then industry consolidation, with perhaps a dozen or even fewer significant players left a decade later. The fast-paced ISP cycle brought many lessons to the telecoms operators but also many of the lessons were then dismissed when the Internet bubble burst.

Fixed telecoms operators faced competition in the long-distance and international voice business but the last mile in local loop unbundling did not bring significant competition. Where fixed telecoms had opportunities created by the regulators, and in some cases faced no technical limits to competition, cellular telecoms were protected by remarkably high barriers of entry. Only a very small number of radio frequencies could be offered for competitors, often limiting the number of players to a handful. And where a fixed long-distance or international voice operator could purchase start-up equipment for amounts measured in the hundreds of thousands of dollars, the practical deployment of a cellular network typically costs hundreds of millions, if not billions, of dollars.

While cellular telecoms did not see heavy competition during the 1990s, the evolution to 3G and its close competitors such as 2.5G, W-LAN and MVNOs, will dramatically change the competitive landscape. Most 3G licence holders belong to global families such as Vodafone, Orange, T-Mobile, Hutchinson, I-Mode, Telefonica, Telecom Italia, etc., groups. The global backing brings global competence centres and facilitates rapid diffusion of competitive information. Success in one country will ever more rapidly be adopted in sister organizations across the global operator's footprint. There will be no hiding.

The key to winning in 3G will not be technology. This will depend on modern marketing, fierce competitivenss and strict attention to profitability. Sadly, most of the current 3G licence holders have spent millions on training their staff to know the technical details and standardization histories of the technology. That was their traditional stronghold to begin with, and it will not help win the war. Remember that the Betamax was technically superior in every way as a video cassette recorder to its VHS counterpart. VHS won. The history of technological innovation is littered with superior technologies that succumbed to inferior ones that had better marketing and competition.

Figure 19.1 Operators must ensure that they stay out of the Zone of Mediocrity area of the Competition Triangle with a clear and well defined business & marketing strategy

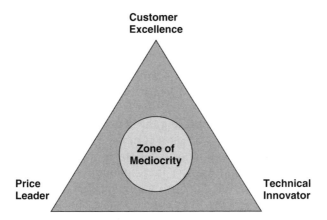

If technology is not the answer, then the answer lies elsewhere. To some degree, winners will be found among those mobile operators who are bold and innovative in creating new services and building customer addiction. Other operators will succeed not by being the ultimate best in service development, but rather in being the best at customer satisfaction. Know your customer and what it takes to keep the customer more happy with you than your competitors. This comes to the heart of marketing. Don't try to be everything to everybody, but focus. To do that well needs segmentation and focus.

The final key to winning is profitability. In the current environment if a company leaks 5% of its revenues per year, no investor or owner will put up with that kind of haemorrhage directly from the bottom line. Some of those left standing victorious at the end will be those with prudent financial controls, and that means effective addressing of revenue assurance. Of course this book is not focused on the financial side of running advanced mobile telecoms business, but the reader must keep in mind that prudent business practice is one of the key determinants in achieving long term success.

The most successful 3G operators — and their close rival competitors such as 2.5G, W-LAN and MVNOS — will not forget the technology, but will focus on the critical management and staff being truly trained in marketing. As we saw in this book, that means starting with customers and then segmentation. Segmentation must govern the company, not technology. Having staff read telecoms marketing books like *3G Marketing* will be a start, but the company will need to go much further, bring in fresh, strong, marketing know-how from industries that know marketing in real competitive situations, such as consumer goods like detergents, confectionary, automobiles, etc. Then if the company is serious about winning, it will need to establish clear management priorities to change from a technology company to a marketing company. This means running marketing training and workshops with leading experts.

As we have stated throughout this book, the future brings ever increasing connectedness. We build and live with our communities, and these communities are increasingly critical of marketing success. As the mobile phone is the defining tool separating the networked generation from the connected generation, those who market in the 2.5G and 3G environment will have to deal with the emergence of strong communities. The early signs are very strong and suggest that communities act as a counterbalance against brands in the way marketing messages are delivered. Anyone involved in 3G marketing must understand communities and learn to work with them, not against them. 3G operators will be in a key position to use this insight in supporting their content and application partners as well.

Where marketing theories and techniques can be trained, a culture of competitiveness is another matter altogether. Again a culture of outperforming the competition is foreign to most established telecoms operators. Competitor insight units seem like afterthoughts. Often their findings are ignored and are rarely linked to management bonuses and incentives. Also, where a company has not done a meaningful analysis of its customer base, random customer opinion samples can actually drive the company in the wrong direction, by focusing on irrelevant issues of marginal customers rather than the pressing needs of the most profitable customers. Segmentation and customer understanding must be at the heart of all competition analysis as well. While competitiveness is not the primary purpose of this book, most marketing of course aims at winning the hearts and minds of the customer. Success in the market place — within target segments — is the natural aim of marketing. Thus this book is also about winning in the new competitive space.

The battle for the customer will be fought in the stores, both owned by operators, and operated by independents; in the calling centres; and increasingly by the content and application partners. It is vital that the front line staff is well trained, has the right attitude of customer service, and is motivated to satisfy the customer. The same is true of the customer-facing staff at the partners. New technologies will never be deployed flawlessly, we have to be prepared to admit and correct mistakes. Customers understand and are willing to give us a chance to fix things. But the staff will need to communicate an image of caring. The market share competition with new mobile telecoms services will increasingly become a contest of execution. Towards the end of this decade the difference between the winners and losers will increasingly come down to better staff and management.

We wrote this book to help marketing managers bring modern marketing methods to mobile telecoms, and to show how certain unique aspects of mobile telecoms can be exploited in marketing. In writing this book we often had to re-examine our own views and reconsider prevailing assumptions. We hope this book has inspired you. As a last thought, we also hope you can bring inspiration to your team and help to build a greater 3G world for all of us. As Anita Roddick said: 'Passion persuades. More than intellectual debate, more than reasoning, more than bloody strategy plans'.

Being Part of the 3G Revolution

Humorous Interlude

You know when you have missed the 3G revolution, when:

- you are the object of constant peer pressure by your family and friends to update your mobile phone;
- you possess only one (1) telephone;
- that annoying blinking sign of an envelope on your mobile phone screen keeps disturbing you, but you don't dare to ask your grandchildren again how to open the text message;
- you still have to CALL home to check what to buy at the grocer's — instead of just being connected to the camera in your fridge;
- you find latest news from a newspaper, watch weather forecasts on TV and listen to traffic updates on the radio;
- you still leave voice mail instead of hanging up at the announcement and sending a text message instead;
- friends dismiss you because you don't have your terminal switched on 24/7;

3G Marketing: Communities and Strategic Partnerships   Tomi T. Ahonen, Timo Kasper and Sara Melkko
© 2004 John Wiley & Sons, Ltd   ISBN: 0-470-85100-7

- you think this list is funny and you want to *fax* it to your friends...(wake up friend, nobody uses a fax any more);
- you think this list is NOT funny and you want to harass your friends by mailing it to them...except you don't have any stamps;
- you keep having to explain that yes you do have a mobile phone and yes it does function, only that you don't see the need to carry it everywhere.

You know when you are part of the 3G revolution, when:

- you have walked into a traffic sign while composing a message or playing a game while walking;
- When you look back the good old times when a terminal was still called a mobile phone and the only thing you could do with it was to telephone;
- your drawers are full of terminals, wires, rechargers, spare batteries, hands-free sets and none are compatible with another;
- whenever you see a vending machine you instinctively point your mobile phone at it;
- you feel like all of society has forgotten you if you have not received messages from anyone for the last two hours;
- the monthly bill to the provider is higher than the rent for you apartment...and you are more than happy to pay it;
- you feel completely lost when you forget your mobile phone, and suicidal when your mobile phone gets stolen;
- you develop a tamagochi-like relationship with your terminal and speak of "feeding the baby" when you recharge the battery;
- you are so accustomed to taking snapshots of everybody with your camera phone that you have accidentally taken a picture of your own reflection in a mirror;
- you believe that there is a secret burial ground for old terminals somewhere in Lapland;
- you say to your spouse "honey, they play our ringing tone."

The above is for entertainment purposes only. It does not reflect the opinions of any of the employers, present or past, nor of current customers, colleagues of the authors, nor of any companies or public personalities mentioned. Original humour by Sara Melkko, Timo Kasper and Tomi T Ahonen/ HatRat 2004.

Abbreviations

1G	First Generation mobile networks
2G	Second Generation mobile networks
2.5G	mobile networks enhanced beyond 2G but that are not 3G
3D	Three Dimensional
3G	Third Generation mobile networks
4G	Fourth Generation mobile networks
5 M's	Movement Moment Me Money and Machines
802.11	IEEE standard for wireless connectivity, see W-LAN and Wi-Fi
802.16	IEEE standard for broadband wireless, see WiMax
802.20	IEEE standard for broadband wireless
ABC	Always Best Connected
AIDA	Attention, Interest, Decision, Action

3G Marketing: Communities and Strategic Partnerships Tomi T. Ahonen, Timo Kasper and Sara Melkko
© 2004 John Wiley & Sons, Ltd   ISBN: 0-470-85100-7

AMPS	Advanced Mobile Phone System (also American Mobile Phone System, also Analog Mobile Phone System)
ARPU	Average Revenue Per User
B2B	Business to Business
B2C	Business to Consumer
B2E	Business to Employee
BBS	Bulletin Board System
CD	Compact Disk
CD-ROM	Compact Disk, Read Only Memory
CDMA	Code-Division Multiple Access
CDMA 2000 1xRTT	Code-Division Multiple Access 1x Radio Transmit Technology
CDMA 2000 EV-DO	Code-Division Multiple Access Evolution Data Only
CDMA 2000 EV-DV	Code-Division Multiple Access Evolution Data and Voice
CDR	Charge Detail Record
CEO	Chief Executive Officer
CRM	Customer Relationship Management
DIN	Deutchse Industrie Norm(ung)
DNA	Deoxyribo Nucleic Acid
DRM	Digital Rights Management
DVD	Digital Video Disk
EDGE	Enhanced Data Rates for GSM Evolution
ERP	Enterprise Resource Planning
GPRS	General Packet Radio Service
GPS	Global Positioning System
GSM	Global System for Mobile communications
Hi-fi	High Fideility
HSCSD	High Speed Circuit Switched Data
HTML	HyperText Markup Language
IMT-2000	Internatinal Mobile Telecommunications 2000
IEEE	Institute of Electrical and Electronics Engineers
IP	Internet Protocol
IR	InfraRed
ISDN	Integrated Services Digital Network
ISP	Internet Service Provider

IT	Information Technology
ITU	International Telecommunications Union
J2EE	Java 2 Enterprise Edition
LAN	Local Area Network
LP	Long Play record
M2M	Machine to Machine
mAd	mobile Advertising
m-Commerce	mobile commerce
m-DJ	mobile DJ Disk Jockey
MMS	Multimedia Messaging Service
MOU	Minutes Of Use
MP3	MPEG-2 Layer 3 (Motion Picture Experts Group)
MSN	MicroSoft Network
MVNOs	Mobiel Virtual Network Operator
NMT	Nordic Mobile Telecom
OFTEL	Office of Telecommunications
P2P	Person to Person
PC	Personal Computer
PDA	Personal Digital Assistant
PDC	Personal Digital Cellular
PIM	Personal Information Manager
PIN	Personal Identification Number
PR	Public Relations
PTT	Post, Telegraph and Telecoms
R&D	Research and Development
ROI	Return On Investment
SIM	Subscriber Identity Module
SME	Small and Medium Enterprise
SMS	Short Message Service
SMSC	Short Message Service Centre
SOHO	Small Office Home Office
SOM	Self-Organizing Map
SVGA	Super Video Graphics Array
SWOT	Strengths Weaknesses Opportunities Threats
TDMA	Time Division Multiple Access
TD-SCDMA	Time Division Synchronous Code Division Multple Access
UMTS	Universal Mobile Telecommunications System

UP	User Plane
USIM	Universal Subscriber Identity Module
VAS	Value Add Service
VCR	Video Cassette Recorder
VHS	Video Home System
VIP	Very Important Person
WAN	Wide Area Network
WAP	Wireless Application Protocol
W-ASP	Wireless Application Service Provider
W-CDMA	Wideband Code Division Multiple Access
Wi-Fi	Wireless Fidelity
WiMax	Worldwide Interoperability for Microwave Access
W-LAN	Wireless Local Area Network
WWW	WorldWide Web
WYDIWYG	What You Demand Is What You Get
XML	eXtensible Markup Language
ZIP	Zone Improvement Plan

Bibliography

Adler, J., Halata, E., Holbert, N. Kunden halten oder Märkte erobern? Economica Hüthig Verlag, Heidelberg, 2001.
Ahonen, T. m-Profits, Chichester: Wiley, 2002, 390 pp.
Ahonen, T., Barrett, J. Services for UMTS, Chichester: Wiley, 2002, 373 pp.
Frengle, N. i-Mode, A Primer. M&T Books, New York, 2002, 485 pp.
Funk, J. Mobile Disruption. Chichester: Wiley, 2004, 211 pp.
Golding, P. Next Generation Wireless Applications. Chichester: Wiley, 2004, 576 pp.
Halonen, T., Melero, J., Romero, J. GSM, GPRS & EDGE Performance. Chichester: Wiley, 2002, 614 pp.
Hannula, I., Linturi, R. 100 phenomena, virtual Helsinki and the cybermole (translation from Finnish by William More) e-book at http://www.linturi.fi/100_phenomena/ Helsinki: Yritysmikrot, 1998 212 pp.
Holma, H., Toskala, A. WCDMA for UMTS, revised edition. Chichester: Wiley, 2001, 313 pp.
IDATE. Web Music: Issues at Stake and Forecasts. Montpellier 2001.
Kaaranen, H., Ahtiainen, A., Laitinen, L., Nahgian, S., Niemi, V. UMTS networks. Chichester: Wiley, 2001, 302 pp.

3G Marketing: Communities and Strategic Partnerships Tomi T. Ahonen, Timo Kasper and Sara Melkko
© 2004 John Wiley & Sons, Ltd   ISBN: 0-470-85100-7

Kahaner, L. Competitive Intelligence. New York. Touchstone – Simon & Schuster Inc., 1997, 300 pp.

Kopomaa, T. City in your pocket, the birth of the information society. Helsinki: Gaudeamus, 2000, 143 pp.

Kotler, P., et al. Principles of Marketing, FT Prentice Hall, 2001, 888 pp.

Lamont, D. The Age of m-Commerce. Oxford: Capstone, 2001, 288 pp.

May, P. Mobile Commerce. Cambridge: Cambridge University Press, 2001, 302 pp.

May, P. The Business of E Commerce. Cambridge: Cambridge University Press, 2000, 288 pp.

McLelland, S. Ultimate Telecom Futures. Guilford: Horizon House, 2002, 232 pp.

Merrill Lynch. Wireless Matrix 3Q202, 10 December 2002.

Moore, G. Crossing the Chasm. Capstone Publishing, 1999, 222 pp.

Ralph, D., Graham, P. MMS Technologies, Usage and Business Models. Chichester: Wiley, 2003, 382 pp.

Rheingold, Howard. Smart Mobs. Perseus Books, 2002, 288 pp.

Sadeh, N. M-Commerce. Chichester: Wiley, 2002, 272 pp.

Schmalen, H. Grundlagen und Probleme der Betriebswirtschaft. Schöffer-Poeschel, 2002, 800 pp.

Tennent, J., Friend, G. Economist guide to business modelling. London: Economist Books, 2001, 272 pp.

UMTS Forum, Report 13: Structuring the service opportunity, available at the UMTS Forum website http://www.umts-forum.org/reports_r.html April 2001.

UMTS Forum, Report 16: 3G portal study, available at the UMTS Forum website http://www.umts-forum.org/reports_r.html November 2001.

UMTS Forum, Report 21: Billing and Payment Views on 3G Business Models, available at the UMTS Forum website http://www.umts-forum.org/reports_r.html July 2002.

UMTS Forum, Report 26: Social Shaping of UMTS, available at the UMTS Forum website http://www.umts-forum.org/reports_r.html January 2003.

Wilkinson, N. Next generation network services. Chichester: Wiley, 2002, 216 pp.

Webb, G. The M-Bomb. Oxford: Capstone, 2001, 256 pp.

Useful Websites

160 characters (SMS and Mobile Messaging Association)
http://www.160characters.com
3GPP (Third Generation Partnership Project)
http://www.3gpp.org
ALACEL (Latin American Wireless Industry Association)
http://www.alacel.com/home.cfm?lang=en
ARIB (Association of Radio Industries and Business) (Japan)
http://www.arib.or.jp/index_English.html
CTIA (Cellular Telecommunications & Internet Association)
http://www.wow-com.com
CWTS (China Wireless Telecommunication Standards Group)
http://www.cwts.org/cwts/index_eng.html
Ecademy
http://www.ecademy.com/
ETSI (European Telecommunications Standards Institute)
http://www.etsi.org
FCC (Federal Communications Commission)
http://www.fcc.gov

3G Marketing: Communities and Strategic Partnerships   Tomi T. Ahonen, Timo Kasper and Sara Melkko
© 2004 John Wiley & Sons, Ltd   ISBN: 0-470-85100-7

GBA (Global Billing Association)
http://www.globalbilling.org
GSA (Global mobile Suppliers Association)
http://www.gsacom.com
GSM Association
http://www.gsmworld.com
IDC
http://www.idc.com
IEEE (Institute of Electrical and Electronics Engineers)
http://www.ieee.org
IETF (Internet Engineering Task Force)
http://www.ietf.org
Informa
http://www.telecoms.com
ITU (International Telecommunications Union)
http://www.itu.ch
John Wiley & Sons
http://www.wiley.com
Jupiter Research
http://www.jup.com
MDA (Mobile Data Association)
http://www.mda-mobiledata.org
MEF (Mobile Entertainment Forum)
http://www.mobileentertainmentforum.org
MGIF (Mobile Gaming Interoperability Forum)
http://www.mgif.org
MMA (Mobile Marketing Association)
http://www.waaglobal.org
PayCircle
http://www.paycircle.org
PCIA (Personal Communications Industry Association)
http://www.pcia.com
PPA (Periodicals Publishing Association) UK
http://www.ppa.co.uk
SyncML
http://www.syncml.org/
The Research Room
http://www.theresearchroom.com
The 3G Portal
http://www.the3Gportal.com
Tilastokeskus / Statistics Finland
http://www.tilastokeskus.fi/index_en.html

TTC (Telecommunication Technology Committee) (Japan)
http://www.ttc.or.jp/e/
UMTS Forum
http://www.umts-forum.org/
WAP Forum
http://www.wapforum.org
WISPA (Wireless Internet Service Providers Association)
http://www.wispa.org/
WLANA (Wireless LAN Association)
http://www.wliaonline.com/
WLIA (Wireless Location Industry Association)
http://www.wliaonline.com/

Index

1G 2, 11
2.5G 6, 10, 20, 55, 110, 145, 259, 296
24 hour news 80
2G 2, 11, 110, 143, 178, 203, 259
3.5G 11
30 minute tasks vs 30 second
   tasks 281
3Com 23, 181, 197
3D rule – Duration, Display,
   Demand 128
3G
  definitions 10
  launch 6, 142
  Launch Strategies (report) 179
  license 237, 296
  standardisation 23
  system 9
  terminal 203
  vs 4G 12
  vs W-LAN 13, 236
3M 160

48 hours, contract management 87
4G 11
5 M's service development 56, 87, 182
802.11 *see* W-LAN
802.20 11

ABC Always Best Connected 109
Abstract product/service 61, 141, 152,
   158, 204, 220
Acceptable price 231
Account manager 208
Addiction 173, 192, 231
Adoption curve 168
Adult
  entertainment 58, 79, 184
  services and sex shops 206
  sex lines 18
Advertising 18, 39, 46, 58, 134, 204,
   211, 288
Advertising *see also* mobile advertising
Age and calling pattern 25

3G Marketing: Communities and Strategic Partnerships Tomi T. Ahonen, Timo Kasper and Sara Melkko
© 2004 John Wiley & Sons, Ltd ISBN: 0-470-85100-7

Agency 288
Ahonen, Tomi 9
AIDA rule Attention Interest Desire Action 140, 203
Airline 38, 175
　analogy 221
　and bullet trains analogy 12
　check-in by mobile 58
Airport 236
Airtime revenue 5
Alarm 25
Alerts 253
Allen, Woody 238
Almes, Guy 189
Alpha users 50
Altavista 126
Amazon 126, 156
America
　brands 156
　penetration 173
American Express 102, 243
AMPS American Mobile Phone System 11
Analogy
　airline seats 221
　airplanes and bullet trains 12
　radar and ships 25
Answer ringing phone 191
Answering machines 176, 191
AOL America On-Line 126
Application developer 72, 85, 146, 211
Arbitrage 243
Argumentation 205
Ariel 152
ARPU Average Revenue Per User 246
Arthur Andersen 147, 158
Asia
　brands 156
　data revenues 3
　revenue share 81
Aston Martin 148

Attention 140
Attorney 86, 89
Attract customers 275
Audi 159
Austria, penetration 173, 269
Authentication 129
Automation 59, 87
Automobile 41, 104

B2B Business to Business 65, 175, 208
Backstreet Boys 4
Bad publicity 28
Bad-will 28, 164
Balance reminder 204
Ballet 148
Banking 211, 242
Banner ads 134
Barrett, Joe 9
Basketball 36, 80, 163
BBC British Broadcasting Company 80
BBS Bulletin Board System 126
Beatles 16
Beeper 173
Behaviour 8, 17, 32, 35, 40
Behaviour and segmentation 41
Bell Canada 273
Benchmark 62, 264, 282
Bendable screen 108
Bentley 159
Betamax 24, 62, 296
Betting 184
Beyond segment of one 47
Billboards 144
Billing 40, 58, 75, 84, 131, 206, 211, 216, 239, 244, 284
Bloomberg 80
Bluetooth 10, 97, 108, 110
BMW 104
Boilerplate 86
Bollinger 148
Bonus and boss 225

Index

Bonus points 277
Books 88
Brainstorming 63
Brand 80, 83, 125, 141, 151, 163, 206, 211, 215, 264, 281
Branding vs communities 166, 298
Brazil subscribers, penetration 7
Broadband 110, 115
Browser 63, 126
Bulk discount 224, 235
Bullet trains and airplanes analogy 12
Bundling 27, 88, 119, 207, 223, 232
Bus tickets 182
Business
  customers 34, 43
  drivers 16
  intelligence 16, 18
  models 253
  plan 288
  strategy 28
  vs residential 48, 174
Butler, Samuel 203

Cable TV news 80
Cadillac 104
Calendar 97, 280, 283
Call detail record 240
Call-to-action 59
Calling
  card 225
  centre 120
  party pays 173
  pattern and age 25
Camera phones 5, 97, 177, 183, 201
Canada
  hockey stick 199
  loyalty points 277
  WAP 6
  water 154
Cannibalisation 3, 68, 285
Capacity 221

Car 2, 59, 91, 104, 190
  industry 41
  racing 148
  racing game 84
Car-wash 4
Carlyle, Thomas 167
Carnegie, Dale 14
Carphone Warehouse 114, 121
Carrier *see* operator
Cartoon 57, 79, 87, 245
Cash machine 56, 243
Cassettes 92
Catch-22 29
CD Compact Diskette 93, 120, 149
CD Player 183, 201
CDMA2000 10
CDMA Code Division Multiple Access 11, 34, 178
CDR Charge Detail Record 240
CD-ROM Compact Disk Read Only Memory 55, 105, 182
Ceiling 79, 172
Celebrities 39
Cell revenue size 53
Cells 51
Cellular *see* mobile
Centre of excellence 28
Champ car racing 84
Change of behaviour 8, 17, 40, 189
Change plans 193
Changer 271
Changing numbers 276
Channel conflict 118
Charge detail record 240
Charging 40, 84, 239
Charlies Angels 2 166
Chasm 141
Chat 127, 184, 207, 245, 283
Chicken and egg 176
China subscribers, penetration 7
Chinese restaurant 129
Churchill, Sir Winston 151

Churn 8, 31, 39, 51, 134, 250, 267
CIA Central Intelligence Agency
    20, 266
Cigarettes 39
Cingular 156
City guide 243
City in the Pocket, The (book) 8
Clandestine 170
Class of service 221
Click to buy 59, 208
Clock 93
Clothing 91, 107
Cluster 32, 231
CNBC 80
CNN 80
Co-branding 160
Coca Cola 4, 151, 166
Coins vs mobile phone 183
Colour screen 5
Commodore 63
Communication 57, 77
Communication of Innovations
    (book) 168
Community 6, 16, 50, 57, 76, 149, 166, 217, 259, 280, 282, 298
Company radar 25
Compaq 97
Competition 10, 21, 31, 155, 297
Competitive assets 28
Competitiveness 16, 31, 221, 298
Competitor analysis/intelligence 16, 20, 298
Competitors and sales force 65
Complaints 249, 272
Compuserve 126
Confidential information 73
Conflict 67, 77, 118, 164
Congestion, car 59, 105
Connectedness 298
Connectivity 57
Conservatives (customers) 142
Consultative sales 66

Consumer rights 18
Consumerism 152
Contact management 86
Content
  aggregators 20, 72, 86
  migration from fixed to mobile 59
  partner 5, 132
  providers 20, 72, 85, 146, 211
  revenue 5, 58, 78
Contract 34
Contract customer *see* postpaid
Controller 88
Convergence 21, 93, 145
Cooper, Martin 2
Copiers and mobiles 59, 107
Copyrights 65, 83, 264
Core business 28
Corporate customers 34, 208
Cosmo Girl MVNO 258
Cost 220, 229, 278
Counter-intuitive 60
Coupons 58, 149, 243
Creating time 57
Credit
  card 58, 84, 102, 183, 191, 224, 242
  risk 211
  services 184
Cricket 80
Cringely, Robert X 111
CRM Customer Relationship
  Management 38, 75, 204, 245, 284
Cross-branding 159
Cross selling 136
Cross-selling 216
Crossing the Chasm (book) 141
Culture 8, 76, 162
Customer 9, 31, 39, 65, 72, 119, 164, 220, 230
  insight/information/intelligence 19, 32, 133, 224, 246, 284
  service 114, 162, 216, 297
  see also subscriber

Index

Customer-perception 226
Customisation 57, 95, 201
Cycle 241

Daimler-Chrysler 104, 118
Data
  from billing 40, 133
  mining 49, 133, 245
Dating 127, 184, 207
Debit cards vs mobile phone 183
Decisions, strategic 17
Deep information on customers 32
Demand and supply 228
Demographics segmentation 36
Demoralised 212
Denmark
  as reference market 27, 191
  cash machines 58
Department store 116
Deployment 66
Design, car industry 46
Desire 141
Developer community 76
Development of services 55
Devices *see* terminals
Diamonds 151
Diesel jeans 155
Digital 3, 82, 220, 264
  camera 97
  content 183
  rights management 83, 244
  TV 182
Dimensions and segmentation 47
Diners Club 243
Discount 241
Discrete segments 32
Disloyal 274
Disney content in Japan 8, 257
Disney, Walt 1
Distribution 38, 113, 224
Distributor 204
DJ Esko 255

Doctors 118
Doonesbury, Mike 125
Dragon 293
Drive-through 190
DRM Digital Rights
   Management 244
DVD Digital Video Disk 24, 93

E-mail 95, 181, 184, 186, 195
Early adopter countries 8, 191
Early adopters 142, 167, 179
Early Eight 182
Early majority 168
Eastern Europe, revenue share 81
Economic cycles 144
Economist 44, 187
EDGE Enhanced Data rates for
   GSM Evolution 11
Educating market 230
Einstein, Albert 137
Elasticity 220, 229
Electricity 240
Electricity bill 146
Electricity meter 59
Elisa 253
Eminem 4, 149
End of service life cycle 68
Engineering
  IT vs telecoms 76
  vs marketing 46
Enron 147, 158
Entertainment 5, 57, 127
Environment scanning 25
Equipment vendor 73, 76, 142
Ericsson 148
  3G phone 94
  ABC Always Best Connected 110
  change of behaviour 191
  and supply of components 26
  see also Sony Ericsson
ERP Enterprise Resource
   Planning 38

Europe
  3G licences 2
  ARPU 246
  brands 156
  content revenue 58
  i-Mode 78
  launch 3G 6
  location based services 185
  mobile content revenue 253
  revenue share 81
  SMS 3, 87
  WAP 6, 132
Evangelists 180, 212, 231, 284
EV-DO Evolution Data Only 6, 10
EV-DV Evolution Data and Voice 11
Evenings vs office hours 43
Everybody wins 29
Evian 154
Evolution of gadgets 92
Evolution of user 192
Excel 63, 99
Exclusive 79
Exclusivity 82, 263
Executives use SMS 57
Exponential curve vs hockey stick 199
Export license 26

Facts 21
Fake IP address 129
Farm animals 174
Fax 23, 107, 176
Ferrari 164
Fiat 104
Filofax 201
Financial resources 58
Financial services 184
Financial Times 118, 246, 258
Financing 26
Finland
  adult entertainment 18
  airline check-in 58
  as reference market 27, 191
  first digital 2
  hockey stick 199
  ISPs 253
  lottery 261
  MVNO 258
  no subsidies 108, 214, 278
  penetration 173, 191, 269
  public transport and mobile 8, 280
  saturation myth 173
  SMS 8, 87
  turn off 1G network 11
Finnish proverb 287
First
  class 222
  game free 262
  impression 67
  to reduce price 227
Fishing game 5
Fishing license 184
Five M's *see* 5 M's service
    development
Fixed internet
  adult entertainment 58
  advertising 59, 144
  as sales channel 117
  comparison 7, 183
  competitor intelligence 20
  and credit cards 191
  emergence 1, 23
  no billing system 58
  penetration, Japan 5
  services 55
  and university 24
  use fixed subscription 169
  vs mobile Internet 135, 182
Fixed telecoms
  operators 20, 169
  price 229
  vs mobile 7, 189
Flash 101
Floor 79
Focus groups 35, 64

Index

Follower   227
FOMA   6, 100
Football   80, 148, 207
Footprint   296
Ford   35, 46, 104, 160
Ford, Henry   69
Forecasting   27, 169
Formula One   84, 207
Four P's Product Place Price
    Promotion   140
France
    early adopters   180
    subscribers   7, 231
Fraud   75, 247
Frederic the Great   295
Free
    calls by employer   45
    internet access   255
    SMS   207, 262, 276
Frequency   296
Friendly user trial 3G   6
Friends and family   232

Gambling   7, 87, 184
Gameboy   101
Gaming   3, 7, 79, 84, 87, 91, 93,
    101, 201, 262, 279
    and car   106
    networked   95, 110
    revenues   183, 217, 220, 253
Gas meter   59
Gates, Bill   15
GE General Electric   153
General Motors   104
Generation   169
Germany
    3G license   237
    airline ticket   223
    car makers   159
    early adopters   180
    hockey stick   199
    penetration   7, 171

SMS advertising   146
    subsidies   214
Getty, J Paul   55
Gift item bundle   27
Global operator   142, 296
Global positioning   84
Google   135
Gossip   88
Government customers   209
GPRS General Packet Radio
    System   6, 11, 34, 120, 177,
    182, 204
GPS Global Positioning Satellite
    84, 95, 103, 105, 110, 259
Grandmothers   8
Greedy   256
Greenfield   61, 179
Growth   31
GSM Global System for Mobile
    communications   2, 11, 34,
    178, 198
Gustafsson, Eva   110

Habits, change   8
Haircut   229
Hairdresser   206
Handheld device   183
Handset subsidies   108, 214
Handsets *see* terminals
Harley Davidson   154, 284
Harmonisation of price   229
Harry Potter   87
Harvesting   62
Harvey, Paul   139
Heineken   135
Heinz   157
Help desk   120, 213
Helsinki   280
Herd behaviour   166
Heterogenous   41
Hi-fi High Fidelity   92
Hidden churn   268

Hierarchy of plans   288
Hiking   206
Hockey Stick vs Metcalfe's Law   198
Hollywood   101, 159
Holma, Harri   9
Home entertainment   93
Home zone   236
Homogenous   41, 152
Hong Kong
  mobile advertising   262
  MVNO   258
  penetration   173, 269
  rapid growth   7
Horoscopes   57, 207
Horseless carriage   190
Hospitals   118
Hot spot   236
Hotel   225, 236
Housewife   43
HSCSD High Speed Circuit Switched Data   11
HTML HyperText Markup Language   5, 257
Hunting license   184
Hutchison   6, 296
Hybrid user   48
Hype   6, 13, 141–4
Hyper growth   7
Hypothetical   52

IBM   116, 153
Ice hockey   80, 148
Iceland, penetration   173
Idea generation   63
Illegal   170
I-mode   5, 11, 78, 85, 87, 182, 255, 263, 296
Impulse purchases   57
IMT-2000 International Mobile Telecoms 2000   11, 141
Incumbent   4, 157
Industrial espionage   20

Industry type segmentation   36
Indy car racing   84
Inflection point   198
Infomercials   145
Information   5, 17
  confidential   73
  deep on customers   32, 285
  and sales force   122
Infoseek   126
Initiate calls   42, 50
Innovation   62
Innovators   142, 168, 284
Insight   17, 32
Insurance   145, 184
Intangible product or service   61, 119, 184
Intel   153
Intellectual Property Rights see IPR
Interaction with machines   185
Interchangeable covers   3, 57, 201, 280
Interconnect   4, 22
Internal marketing   67
International calls   1, 204, 254
Internet see fixed Internet or mobile Internet
Internet Explorer   63, 126
Internet to the pocket   6
Interview   64
Intimate messages   18
Intranet   116
Introductory price   234
Inventory   122
Invoice   241, 246
Iobox   233
IP Internet Protocol   24, 129
IPR Intellectual Property Rights   65, 183
IR InfraRed   11, 97, 110
ISDN Integrated Services Digital Network   23
Isle of Man, 3G   101

Index   321

ISP Internet Service Provider   4, 75, 126, 252, 295
Israel, penetration   8, 173, 269
IT Information Technology department   209
IT integrators   20, 72, 116, 204, 211
Italy
  3G   6
  early adopting country   191
  loyalty points   277
  no subsidies   108, 214, 278
  penetration   7, 171, 173, 231, 269
  prepaid   3, 169
  SMS advertising   146
Itemised billing   241
ITU International Telecommunications Union   13

J2EE Java 2 Platform Enterprise Edition   24
Jaguar   104, 159
James Bond   148
Japan
  3G   6
  airline tickets by mobile   58
  ARPU   246
  brands   156
  camera phones   177
  click-to-buy   208
  Disney   8
  entertainment   57
  first cellular   2
  fixed internet penetration   5
  I-Mode   5, 78, 85
  itemised billing   241
  messaging   8, 58
  penetration   7
  portals   131
  reference market   27
  revenue sharing   5, 78, 81, 85
  trains and mobile   8

  translation   59, 185
  WAP   6
Java   95
Jet airliners   175
Jippii   4, 253
Joiner   269
Jones, Steve   179
Jonsson, Annika   110
Journalist   143
J-Phone   5, 58, 156, 177, 255
Junk mail and spam   19, 28, 162, 285

KDDI   5, 59, 185, 255
Keyboard   97
Keystroke   83, 264
Killer application   3, 186, 195, 245, 254, 281
Killer cocktail   186
Killer technology   144
Killing a service   68
King, Stephen   93
Kiss MVNO   258
Know your customer   32
Knowledge   17
Kolumbus   253
Kopomaa, Timo   8
Korea
  3G   6
  brands   156
  click-to-buy   208
  credit cards and mobile   103
  no subsidies   109, 278
  portals   131
  reference market   27
  subscribers, penetration   7
  WAP   6

Laggards   168
Laiho, Wacker & Novosad   9, 123
Lamborghini   159
LAN Local Area Network   23, 116

Laptop PC   97, 103, 117, 237
Late majority   168
Latin America   84, 173
Launch   67, 167
Law of supply and demand   228
Lawsuits of wrong bill   248
LBS Location Based Services   184
Leakage   247
Learning curve   76
Leaver   270
LeCarre, John   166
Legacy IT   177
Legal restrictions   65
Legal system   18
Leno, Jay   250
Letterman, David   266
Levi's jeans   153
LG   215
Liberalisation   295
License   2, 184
Life cycle   62, 68
Lifespan   88
Lincoln   104, 160
Link and price shown   236
LM Ericsson *see* Ericsson and Sony Ericsson
Local calls   254
Local sports   80
Location   56, 82, 84, 184, 211, 259, 264
Lock customers   275
Locksmith   120, 206
Logos, downloadable   4, 132, 217, 254, 280
Long distance calls   1, 232, 254
Look and feel   155
Lord of the Rings   87
Loss-leader   88
Loss-making customer   271
Lost profit   229
Lost time   57
Lottery   184, 261

Lotus 1–2–3   63
Love notes and SMS   43
Loyalty   39, 136, 155, 271, 277
Lycos   126

M-banking   79
M-Commerce   11, 45, 136, 204, 235, 249, 261
M-commerce partners   211, 251
M-DJ mobile DJ   235
M-Profits (book)   9
McDonalds   135, 153, 243, 262
Machine subscriptions   172
Machine-to-machine   59
MAd *see* mobile advertising
Madonna   4, 82
Magazines   144
Magic of micropayments   232
Mainstream market   142
Man-machine   59
Management of services   55
Management theory   16
Managing time   57
Manufacturers   20, 143
Manx Telecom   101
Mapping   52, 59, 84, 104, 243, 259
Market
  distortions   278
  intelligence   16, 282
  research   16, 230, 282
  rules   86
  share   31, 249
  view and sales force   65
Marketing
  department   163
  internal   67
  manager   285, 299
  mix   140, 274
  plan   140, 287
  viral   149, 244
  vs engineering   46

Marlboro 153
Marx, Groucho 219
Mass-customised 210
Mass-market 8, 141, 181, 235, 237, 254
Mass-marketing 206
Mastercard 102, 243
Matrix segmentation 51
Maximise profit of service 61
Me (attribute) 57
Media 93
Medical services 184
Megabyte tariffing 235
Member get member 283
Mercedes-Benz 104, 153, 158
Merchandising 88
Messaging 82
Metcalfe, Robert 23, 181, 197
Metcalfe's Law 23, 181, 197
Metering 59
Mexico penetration 7
Microbilling 254
Microchip 93
Micropayment 58, 82, 85, 182, 232, 240, 264
Microprocessor 96, 106
Microsegment 48
Microsoft 63, 95, 126, 153
Microwave 175
Middle East and USA export license 26
Middleware 75
MIDI Musical Instrument Digital Interface 5
Migrating content 59
Milk 229
Minidisk 93, 183
Minimum level 79
Minutes of Use MOU 228
Mistrust 78
MMS multimedia messaging 7, 40, 45, 50, 58, 99, 120, 207

Mobile
  advertising 79, 134, 144, 204, 243, 249, 251, 253, 261
  banking 79, 249
  commerce 5, 11, 136, 204, 251, 261
Mobile Internet
  brands 156
  competitor intelligence 20
  invention 4, 76
  itemised billing 241
  still too young 8, 76
  vs fixed Internet 135
Mobile operator *see* operator
Mobile phones *see* terminals
Mobile telecoms vs fixed 7, 189
Mobility 56, 116
Model T Ford 46
Modem 23, 103, 115, 142, 177
Moment 57, 145
Monaco, 3G car 104
Money (attribute) 58, 146
Monitoring 172
Monopoly 19, 76, 114, 155, 220, 251, 295
Moore, Geoffrey 141
Mosaic 63
Motivate staff 115
Motorola 2, 156, 170, 215
MOU Minutes of Use 228
Movement 56
Movie 79, 262
Moving car 82
MP3 player 93, 183
MTV Music TV 118, 258
Multi-access portal 136
Multimedia 93
Multimedia messaging *see* MMS
Multiple subscriptions 51, 169
Multitasking 57, 106
Music 82, 92, 149
  industry 4, 117
  player 97, 201

MVNO Mobile Virtual Network
    Operators   20, 75, 86, 115,
    118, 211, 234, 258, 274, 296
Myth of near saturation   173

Narvesan MVNO   258
NASCAR   84
Navigation   104, 127, 201, 259
Near saturation myth   173
NEC   100, 156
Nelly MVNO   258
Netscape   63
NetWare   63
Network   23, 27, 66
Network hubs, alpha users   50
Network selling   217
Networked gaming   95, 110
Neural network   48
New England Journal of
    Medicine   250
New services   31, 62
New York   190
News   88, 207
Newspapers   144
Newsstand   206
Niche   1, 18, 67, 254, 275
Nike   135, 153
Nintendo   101, 106
NMT Nordic Mobile
    Telecoms   2, 11
Nokia   3, 148, 153, 156, 215
    change of behaviour   191
    ringing tones   255
    segmentation   44
    supply of components   26
Nokia Communicator   98
Nokia N-Gage   102
Nokia Remote Camera   103
North America   84, 283
Northern Europe   184
Norway
    as reference market   27, 191

MVNO   258
parking   4
Notebook PC *see* laptop PC
Novell Netware   63
NTT DoCoMo   2, 5, 78, 85, 87,
    100, 156, 256, 265
Number portability   268, 276

O2   22
Official sites   131
Official websites i-Mode   256
OFTEL   227
Oligopoly   155, 252
Omegas   51
Omnitel Pronto Italy OPI   277
On-line branding   161
On-screen   83, 264
On-Star   104
One bill   246
One price for all   234
Opera   148
Operator
    billing   239
    brand   155
    business   27
    global   142, 296
    greedy   256
    greenfield   61, 179
    growth   7
    handset subsidies   108, 214
    hiring staff   88
    innovation   75
    left alone   1
    loyalty   267
    more competitors   234, 252
    multiple subscriptions   169
    partnership   74, 211, 252
    revenue share   5, 78, 85
    roaming   28
    role   74, 252
    segmentation   33
    services   61, 186

Index 325

stores 114
subscribers 171, 231
tariffing 220, 225
OPI Omnitel Pronto Italy 277
Opinion surveys 4
Opportunity cost 220
Opt in 134
Optimal adoption curve 181
Optimal tariffing 31, 231, 275
Oracle 159
Orange 22, 296
Outsourcing 28

PABX Private Automated Branch
 exchange 284
Packaging 204
Packet based 6
Pager 93, 173
Pain threshold 220, 237, 275
Palmtop computer 98
Panasonic 156
Paper clip industry 240
Paradigm shifts 10
Parents 154
Parking 4, 182
Partnering 27, 71, 86, 132,
 211, 251, 277
Pasteur, Louis 202
Patent 25, 65
Paulson, Hugo 54
Pay-per-game 102
Payback period 209
PC Personal Computer 23, 97, 184,
 201, 296
PC *see also* laptop
PDA Personal Digital Assistant
 91, 95, 117, 177, 201, 262
PDC 11
Peer pressure 283
Penetration rate 1, 7, 168, 198, 269
Pepsi Cola 157, 281
Perception of price 226

Perfect competition 155
Permission based 262
Perrier 154
Personal 184, 191
Personal data segmentation 38
Personalisation 57
Petrol station 190
Pets 174
Philippines
 divorce and SMS 58
 entertainment 57
 political activism 166
 revenue from SMS 3
Phone bill 40, 115
Photocopier 92
Picasso, Pablo 286
Pickup truck 163
Picture messaging 7, 58
PIM Personal Information
 Manager 98
PIN Personal Identification
 Number 129
Pioneer 62
Pizza Hut 56, 243
Plans, hierarchy 288
Playstation 101, 106, 279
Points, loyalty programme 40
Polls 4
Polyphonic 5
Pop Idol 87
Pop star 120
Popular services 255
Population 168, 269
Porsche 155
Portals 20, 75, 86, 118, 125, 130,
 136, 156, 207, 211
Portugal, prepaid 169
Positioning 84, 130
Post-paid 34, 43, 172
Postpaid vs pre-paid price 229
Postponing time 57
PR agency 144, 288

PR public relations 141, 147
Pragmatists 142
Predictive 39
Preece, Sir William 91
Prefer SMS vs voice 42
Premium rate, adult entertainment 18
Premium SMS 73
Prepaid 3, 34, 169, 172, 229, 242, 268
Press plan 144
Press relations 67
Price *see* tariffing
PriceWaterhouseCoopers 247
Primary source 21
Printing house 34
Prisoner's dilemma 109, 215, 280
Privacy 196, 285
Procter and Gamble 160
Product *see* service
Product development cycle 75
Product placement 148
Profile 39
Profit, maximise 61, 226, 232
Profitability 1, 31, 51, 66, 72, 88, 208, 220, 248, 271, 297
Project manager 64
Projection display 108
Promotion 139, 288
Protectionist 255
Provisioning 75
Psychologists 48
Public domain 20, 25
Public relations 143
Public transportation and mobile phones 8
Pull 130, 259
Pure profit 232
Push 259

Quake 101
Questionnaire 40

Racing car game 84
Radar and ships analogy 25
Radar and SOM 51
Radio 95, 131, 144
Radio Network Planning and Optimisation (book) 9, 123
Radiolinja 2, 254
Raw material 26
Razor business 279
Reachability 189
Receive calls 42
Receiving party pays 173, 283
Reebok 153
Reference market studies 27
Refrigerator 107
Regan, Roger 239
Relationships and SMS 58
Remote
  camera 107, 119
  control 172, 185
  payment 172
Rent mobile phone 225
Replace phone 278
Reporting 240
Residential customers 34, 43, 174
Return calls 43
Revenue
  by cell of model 53
  content 58
  music 117
  per minute 22
  split 246, 251
Revenue assurance 247
Revenue leakage 247
Revenue sharing 5, 78, 85, 251
Revenue, SMS 3
Ringing tones 3, 57, 120, 132, 217, 253, 280
Risk 79
Roadmaps 24, 75
Roads 108
Roaming 2

Index

Robertson, James A 149
Robotic cameras 91
Rockefeller, John D 89
Roddick, Anita 299
Rogers and Shoemaker 168
ROI Return on Investment 209
Role 74, 252
Rolls Royce 111, 164
Rules of thumb 84, 265

S-curve 168
Sales 66, 73, 114, 136, 140, 205, 208
   force 39, 46, 65, 122, 204, 212
   to partners 211, 259
Samsung 156, 215
SAP 116
Satellite news 80
Satisfy customers 31, 272
Saturation 169, 172
Saunalahden Serveri 4, 253
Scalability 75
Scandinavia
   cosmetics 27
   growth rate 7
   revenue share 81
   roaming 1981 2
   SMS 57, 200
   telecoms 27
Sceptics 142
Search engine 127, 135, 201
Seat 159
Second phone 43, 283
Secondary source 21
Secretive messages SMS 3, 57
Security 120, 145
Security risk 242
Segment 32, 51, 88, 206, 216, 227, 231, 245, 275
   marketing manager 285
   marketing plan 289
   of one 44

Segmentation 31, 36, 45, 282, 289, 297
   adopters 179
   dimensions 47
   ERP and CRM 38
   model 32, 44, 51
   pay more 221
   powerful 41, 224
   simplistic 208
   SOM 48
   tariffing 231
Self Organising Map SOM 48
Selling boats 36
Selling mobile services 114, 203
Service
   adoption 167
   creation, partnership 76
   creation/design, car industry 46
   development 55, 76
   development, 5 M's 56
   life cycle 62, 68
   management 55, 61
   managers 61, 66, 211, 285
   portfolio 66, 282
   profit maximising 61
   proposition, value-add 78
   provider partnerships 28
   termination (killing it) 68
Services for UMTS (book) 9
Services, fixed Internet 55
Services, new ideas 62
Sex phone lines 18
Sex shops 206
Sharing SMS 58
Shipping 122
Ships and radar analogy 25
Shoes 108
Siemens 215
SIM Subscriber Identity Module 113, 129, 172, 269, 275
Simultaneous 171
Sinclair 63

Singapore   225
Situation and segmentation   40
SK Telecom   156
Skateboarding   206
Skilled labour   26
Smart mobs   166
Smart phones   95
SME Small and Medium
   Enterprise   34, 209
Smith, Adam   71
SMS Short Message Service   3, 11, 34,
   40, 93, 117, 129, 280
   adult entertainment   18
   advertising   144
   boys vs girls   195
   busy schedules   185
   counter-intuitive   60
   divorce   58
   evolution to MMS   99
   first by young   176
   free   207, 262, 276
   hockey stick   198
   inflection point   200
   initiate   50
   killer application   3, 195
   love notes   43, 57
   Metcalfe's Law   181
   premium   73
   pricing   238
   selling   204
   sharing   58
   use   45, 57, 177
   vs email   195
   vs voice   42
   while talking   171
Snowboarding   206
Sociologists   8, 48
Soft launch 3G   6
SOHO Small Office Home Office
   34, 39, 210
SOM Self Organising Map
   48, 51, 245

Sonera   253
Songs   87
Sony Betamax   24
Sony Ericsson   95, 215
Sony Playstation   101, 106
Sony Walkman   153
Soviet Union   19
Spain
   early adopters   180
   penetration   7
Spam   19, 28, 162, 285
Speakerphones   176
Speech recognition   185
Speech synthesis   185
Speed 3G and W-LAN   13
Speedy Tomato   255
Spice Girls   4
Spies   20
Sponsorship   58, 148, 216, 243, 251
Sports   80, 127, 185
Sports cars   39
Spreadsheet   63
Sprint   156
Standardisation   13, 23, 77
Standards and innovation   22
Standing in line   184
Stayer   272
Stealing customers   31
Stereo   92
Stock quotes   185
Stockmann's MVNO   258
Store   114, 206, 213
Strategic
   decisions   17
   partners   28
   plan   288
Streaming   79
Street lamps   92, 108
Stress   67, 72
Stylus   97
Sub-branding   159
Subscriber number   17

Subscribers 2, 169, 240
Subscription 113, 204
Subsidies 108, 214, 269, 278
Success in market share 31
Super Stable MVNO 258
Supermarket 206
Supplier 72
Supply and demand 228
Surveys 4, 35
Survivor Island 87
SVGA Super Video Graphics
   Array 128
Sweden
   as reference market 2, 191
   early adopters 180
   hockey stick 199
   MVNO 258
   penetration 171, 173
   saturation myth 173
   water 154
Swimsuit 184
Swiss army knife 95
Switch networks 8
Switching centre 23
Switzerland, water 154
SWOT Strengths Weaknesses
   Opportunities Threats
   analysis 24
Synchronisation 110
System Porsche 159
Systematic market intelligence 17
Systems beyond IMT-2000 13

Taiwan, penetration 7, 173
Talking machines 59
Tamagotchi 88, 101
Targeting 32, 88
Tariff modelling 233
Tariffing
   churn 269, 274
   fixed telecom 229
   harmonisation 229
   hot spots 236
   introductory 234
   manager 234
   megabyte 235
   one price 235
   optimal 31, 219, 229, 275
   profits 220
   segmentation 231
   trials 234
Tariffs, fluctuation 226
Tax 169
Taxi company 34
TDMA Time Division Multiple
   Access 11
Technical
   innovator 297
   roadmaps 24
Techno-geek 179
Technology
   driven 62, 141
   enthusiasts 142
   hype 13
Teenager 43, 154
Telecom Italia Mobile (TIM) 3, 296
Telefonica 296
Telematics 174
Telephone 23
Telephone vs telegraph 190
Telia 172, 255
Tennis shoes 153
Terminals 91, 142, 178, 201, 204,
   214, 254, 261, 278
Terminating a service 68
Terminology, IT vs telecoms 77
Text messaging *see* SMS
Theory, management 16
Thin client 24
Three-dimensional segmentation 45
Thumb Tribe 5
Ticketing 172

TIM Telecom Italia Mobile   3, 296
Time   57
Time Magazine   24
T-Mobile   22, 296
Tolkien, J.R.R   293
Tolls   182
Tomi Ahonen Consulting   81
Tomlin, Lily   113
Top-up card   116
Toskala, Antti   9
Tourist   59
Tourist class   222
Toyota   104
Tracy, Brian   123
Traffic congestion (car)   264
Training   122, 205, 213, 298
Trains and mobile phones   8
Translation service   5, 59
Transmission revenue   79
Transportation infrastructure   26
Travelling   127
Trial and error   59, 64, 76, 263
Trial tariffing   230, 234
Trillion dollars   14, 80
Trivia game   59
Truth in pricing   226
Try it for free   234
TV   144, 170, 201, 262
TV services (programmes)   55
Twins MVNO   258
Two-dimensional segmentation   45
Two phones   5, 269

UK
   3G   6, 237
   early adopters   180
   mobile phone ownership
      by age   273
   MVNO   118, 258
   penetration   7, 171, 231
   prices   227

replace phone   278
revenue per minute   22
SMS   3, 57, 146, 194
stores   114
street lights   108
UMTS Forum   127, 244
UMTS Universal Mobile
   Telecommunications System
   9, 141
Unilever   160
Unofficial sites   131
Unprofitable service   68
Upgrade   9
Urgent   57, 185
USA
   airline ticket   223
   ARPU   246
   brands   156
   car navigation   104
   early adopters   180
   export license   26
   gaming age   102
   government and GPS   103, 259
   i-Mode   78
   invented cellular phone   2
   mobile data vertical markets   37
   MVNO   258
   not hockey stick   200
   revenue share   81
   SMS   87, 200
   subscribers, penetration   7, 173
   WAP   6
Usage and price   228
Useful segmentation   41
Users *see* subscribers
USIM UMTS Subscriber Identity
   Module   275
Utility of network   23

Value-add   4, 78
Value Add Services (VAS)   11

Index

Value chain 28, 80, 252, 263
Value-creation chain 28
Vane-Tempest, Merja 210
VAS Value Add Services 11
VCR Video Cassette Recorder
    24, 62, 93, 170
Vending 172, 182, 280
Vendor 72, 76, 142
Verizon 156
Vertical 29, 37
VHS 24, 62, 296
Video calls 176, 234
Video clips 87, 207
VIP Very Important Person
    40, 144, 211
Viral 149, 217, 244
Virgin Mobile 118
Virtual operator *see* MVNO
Virtuous cycle 5
Visa 102, 191, 243
Visionaries 142
Vodafone 5, 22, 296
Voice 82
    call and SMS simultaneously 171
    calls, initiate 50
    mail 171, 186
    telegraph 190
    vs SMS preference 42
Voicemail 40
Volkswagen 104, 159, 163
Volvo 155
Voucher 34, 242
Voucher customer *see* pre-paid

Walkman 153, 183
Wall Street Journal 183
Walled garden 131, 275
WAN Wide Area Network 23
WAP Wireless Application
    Protocol 6, 11, 26, 34, 43,
    61, 117, 119, 131, 145,
    177, 182, 204, 238

Warehousing 122
W-ASP Wireless Application Service
    Provider 20
Waterloo 164
Waterproof 95
WCDMA for UMTS (book) 9
W-CDMA Wideband Code Division
    Multiple Access 10, 141
Weakest Link 87
Weather 127
Web browser 24
Western Europe
    revenue share 81
    saturation myth 172
Who wants to be
    a Millionaire 87
Wholesale 38, 224
Wi-Fi *see* W-LAN
Wilde, Oscar 147
Wilson, Wilde 30
WiMax 10
Windows NT 63
Wireless services 11
W-LAN Wireless Local Area
    Network 10, 12, 82, 95, 97,
    103, 110, 177, 236, 296
Woodruff, Jeff 218
Word 63
Word-of-mouth 132, 140, 227
Word processing 63
WordPerfect 63
World Cup football 57
World Radio Conference 13
Worldcom 147, 158
WRC World Radio Conference 13
Wright, Steven 251
Wristwatch 57, 108
WYDIWYG What You Demand is
    What You Get 128

X-Box 101, 279
Xtract Ltd 49

Yahoo! 126
Yankee Group 185
Yellow pages 131, 145, 195
Youth 3, 192, 254
Yves Rocher 27

Zed 253
Zero population 53
ZIP code 36, 84
Zone of mediocrity 297
Zoom 101
Zulu 31

Other books by Tomi T Ahonen:

m-Profits: Making Money from 3G Services
(360 pages, Hardback, John Wiley & Sons, Ltd, 2002)
ISBN 0-470-84775-1
World's first business book next generation wireless, became world's bestselling 3G book October and December 2003. Covers revenues, revenue-sharing, pricing, profits, mobile services. m-Profits includes 170 service ideas including 50 real services in use around the world. Written with a clear money focus, m-Profits includes mobile industry issues as Money Migration, Hockey Sticks, and the 5 M's theory. The book covers service creation, revenue sharing, content partnerships, telecoms economics, and includes contrasts among all major wireless technologies.
Only book to give comprehensive view of the issues of marketing and revenue.

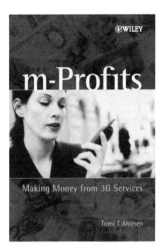

—Sophie Ghnassia France Telecom
Makes complex theories easy to understand using practical examples.
—Mark Weisleder Bell Canada
Down to earth book and great guide to making money in mobile services.
—Joao Baptista Mercer Management Consulting
Good read for industry professionals, operators, bankers and analysts.
—Voytek Siewierski NTT DoCoMo

Services for UMTS: Creating Killer Applications in 3G
Edited by Tomi T Ahonen and Joe Barrett
(373 pages, Hardback, John Wiley & Sons, Ltd, 2002)
ISBN 0-471-48550-0
—*has just been translated to Chinese*
World's first book on 3G services was also the world's bestselling 3G book in October 2002. Covers 212 service ideas with lots of illustrations, statistics, charts and forecasts. Written by 14 leading 3G experts, for the non-technical reader. Includes the 5 M's theory, service creation, content partnerships, revenue sharing, marketing and competition in 3G. Covers all major service groupings such as SMS, MMS, m-commerce, mobile advertising, video, music, gaming, infotainment, B2B, B2C, B2E, etc.
A must read if you want to understand the options and future services.

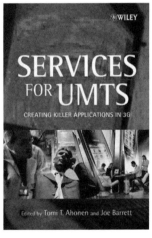

—Dr Roberto Saracco Telecom Italia Future Lab
Significant revenue opportunities to bring value to mobile operators.
—Dr Stanley Chia, Vodafone
Helps operators, manufacturers and capital markets manage $1 trillion bet.
—Assaad Razzouk, Nomura International
Explains some of the compelling services the industry will be able to deploy.
—Jeff Lawrence Intel

Where previous books in this series. *Services for UMTS* **explained the "What", and** *m-Profits* **the "Why" of 3G, this book explains the critical "How". It is a must-read for any professionals involved in making 3G a success.**